THE SOVIET
SPACE PROGRAM

FIRST STEPS: 1941–1953

EBERHARD RÖDEL

Schiffer Publishing Ltd

4880 Lower Valley Road • Atglen, PA 19310

Other Schiffer Books on Related Subjects:
Project Apollo: The Moon Landings, 1968–1972, Eugen Reichl (978-0-7643-5375-8)
Project Apollo: The Early Years, 1960–67, Eugen Reichl (978-0-7643-5174-7)
Project Mercury, Eugen Reichl (978-0-7643-5069-6)
Project Gemini, Eugen Reichl (978-0-7643-5070-2)

Originally published as *Projekt Sputnik: Der Aufcruch ins All* by Motorbuch Verlag,
Stuttgart, Germany © 2014 Motorbuch Verlag, Stuttgart, Germany. www.paul-pietsch-verlage.de

Translated from the German by David Johnston

Library of Congress Control Number: 2017955750

Cover design by Molly Shields

Type set in Avenir LT Std (OTF)/Univers LT 47 CondensedLt

ISBN: 978-0-7643-5539-4
Printed in China

Published by Schiffer Publishing, Ltd.
4880 Lower Valley Road
Atglen, PA 19310
Phone: (610) 593-1777; Fax: (610) 593-2002
E-mail: Info@schifferbooks.com
Web: www.schifferbooks.com

For our complete selection of fine books on this and related subjects, please visit our
website at www.schifferbooks.com. You may also write for a free catalog.

Schiffer Publishing's titles are available at special discounts for bulk purchases
for sales promotions or premiums. Special editions, including personalized covers,
corporate imprints, and excerpts, can be created in large quantities for special needs.
For more information, contact the publisher.

We are always looking for people to write books on new and related subjects. If you
have an idea for a book, please contact us at proposals@schifferbooks.com.

CONTENTS

INTRODUCTION

The reader of this book will have to live with three things: with photos that do not exist ("In keeping with security regulations, photography is forbidden!"), with photographs of poor quality (limited depth of field, poor resolution, because photo technology, like all others in the Soviet Union, lagged years behind the most advanced technologies), and with black-and-white photographs, because the necessary color film was not available. What remain are colorized photos and films. Photographs taken at historical sites cannot make up for this loss of information.

On October 4, 1957, the world held its breath. The beep-beep sound of the first Soviet Sputnik could be heard around the globe. This event caused "Sputnik shock" in the USA, but internationally as well the question arose: where did the technical, material, and personnel potential of the Soviet Union come from?

The basic idea behind this book was to describe the development of Soviet spaceflight from 1945, until the Sputnik era. Naming the significant events and depicting the launch of Sputnik 1 as a victory for Soviet science and technology simply appeared to limit itself to twelve themes.

As I studied the events more closely—beginning with the in-depth study of the German A-4 missile, the creation of the central factory at Bleicherode, and cooperation between Soviet and German rocket experts—it slowly became clear: the complex problems could not be described in a book this size.

Many people still accept the launch of Sputnik 1 as a matter of course. But how was it possible that a nation—which though victorious at the end of the Second World War, had been almost destroyed—was able to put an artificial satellite into orbit before the United States, a highly-industrialized state?

The point of view of that time must also be broadened. The number of questions was considerable. What was the level of Soviet rocket technology in 1945? Where did the many rocket experts come from in a short time? How was such a project even possible under the Stalinist terror apparatus? How could they build devices for which neither the material nor technical conditions existed? Where could they launch rockets? How could they be creative in conditions of enhanced secrecy? Did construction of the R-1 rocket make sense?

These and other questions had to be answered. Contradictory information in the sources and outmoded technical terms and units of measurement were not helpful. Despite all my efforts, I was unable to find any information about many subjects.

So, in the book, I trace the development of the Soviet copy of the A-4—the R-1 rocket, the R2 and R3 rockets, construction of the central state launch complex at Kapustin Yar, to the fact that in 1952, the Soviet rocket industry proved incapable of achieving its primary task—development of a nuclear-armed intercontinental ballistic missile with a range of 6,200 miles. Subsequent development of the R-5 and R-7 rockets, the launch of Sputnik 1, and other satellites will be reserved for later volumes. The origins of the indices (for example 8Ж38 – 8Sh38) also had to be determined. Why did Baikonur have to be built? Why did the USA underestimate the launch of the R-7? Did Sputnik 1 really change our lives? A number of questions thus remain unanswered.

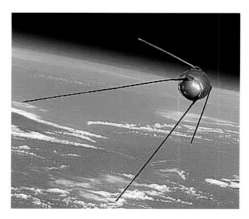

Artist's portrayal of the first Sputnik.

One finds the first answer in the minutes of a meeting held on May 10, 1897: the formulation of the basic rocket equation by Ziolkowski:

$$v(t) = v_g \cdot \ln\left(\frac{m(0)}{m(t)}\right)$$

In 1903, the magazine *Scientific Review* published his principle work "Exploration of Space by Means of Reaction Apparatuses."

GIRD-09 rocket, August 1933; designer M.K. Tikhonravov.

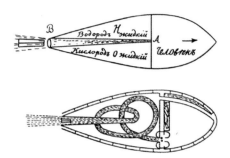

Rocket design drawing from the years 1903 (top) and 1914 (bottom).

In the 1920s, the publication and promulgation of Ziolkowski's writings led to the formation of groups of enthusiastic spaceflight proponents who dedicated themselves to the study of reactive technology. Among them were many enthusiasts who went on to become outstanding scientists in the Soviet space program: Glushko, Langemak, Korolev, and others.

The GIRD (Group for the Investigation of Reactive Propulsion) was established in Moscow and Leningrad in 1931, and later also in other locations. The group's first rocket, the GIRD-09, was launched in March 1933, and reached a height of 1,300 feet.

There were considerable developments in the field of solid- and liquid-fuel rockets and a proven cadre of scientists and engineers. One can assume that by the start of the war there were approxi-mately 700 scientists and technicians in the GIRD and other installations engaged in the development of rocket technology.

The development of rocket-powered aircraft began in 1939, with the I-302, Malyutka, and BI-1. The latter was developed in the Bolkhovitinov Design Bureau (OKB) by Alexander Bereznyak and Alexei Isayev. The abbreviation BI stood for Blizhni Istrebitel (short-range fighter). Disaster struck during the aircraft's seventh flight on March 27, 1943, when it went out of control, killing pilot Grigori Bakhchivandzhi. Despite an intensive search for the problem, the cause of the sudden dive could not be eliminated. Testing was ended.

BI-1 Rocket-Powered Aircraft	
Manufacturer	OKB Bolkhovitinov
First Flight	May 15, 1942
Service Introduction	
Production Period	1941–1943
Number Produced	8

Alexey Mikhailovich Isayev.

Alexey Mikhailovich Isayev was born in St. Petersburg on October 11, 1908, the son of an assistant university professor. In 1925, he began studying at the Moscow Mining Institute. After completing his studies, in 1932, he began working at the country's big construction projects and in project organizations.

He developed a growing interest in aircraft and aviation. In August 1932, Isayev took the post of director of Aviation Factory 22. Two months later he moved to the Bolkhovitinov design bureau outside Moscow. His first project was designing the undercarriage for a long-range bomber. Together with A.J. Bereznyakom, Isayev designed and built the BI-1 rocket-powered fighter (1942). He developed a liquid-fuel engine for the aircraft. During a successful three-minute flight on May 15, 1942, the BI-1 achieved a speed of 250 miles per hour and a height of 8,800 feet. On March 27, 1943, a BI-2 achieved 615 miles per hour, an unofficial world record. After a crash the BI program was cancelled.

In 1944, he became chief designer of a design bureau. In mid-1945, he studied German rocket technology in Thuringia.

In 1946, for the first time in a liquid-fuel rocket engine, Isayev used flat heads with single-component nozzles in a chessboard arrangement. This allowed a high density in burning of the fuel. This head design was widely used in rocket engine design.

In 1947, Isayev became director of OKB-2's development bureau of the NII-1 research institute in Moscow, which was involved with rocket technology for aviation. By the end of the 1940s, the Isayev department developed power plants for surface-to-air missiles and air-to-sea missiles.

The 1950s and 1960s were Isayev's most productive period. His professional partnership with chief designer Sergey Korolev began in 1952. Isayev soon developed an engine for Korolev's R-11 rocket, better known as the Scud and still in service. By 1954, Isayev had also designed power plants for surface-to-air missiles and the Burya missile.

From 1958, he worked in the KB KhimMash. Developed under Isayev were engine systems and micro rocket engines for controlling the majority of Soviet manned and unmanned spacecraft, including the spacecraft Vostok, Vozhkod, Soyuz, Progress, the maneuvering satellites Polyot, Kosmos, the Molnya communications satellite, the space stations Salut and Mir, and for the unmanned lunar and interplanetary stations, oxygen-hydrogen engines for the upper stages of booster rockets, and also engines for the navy's ballistic missiles. Isayev died in Moscow on June 25, 1971.

Doctor of Technical sciences (1959), member of the Communist Party of the Soviet Union (1951).

Honors:
Hero of Socialist Labor (1956), Lenin Prize (1958), State Prize of the USSR (1948, 1968)

Viktor Fyodorovich Bolkhovitinov.

Grigori Bakhchivandzhi, pilot of the first rocket-powered aircraft.

What Is Heroism?
Alexander Matveyevich Matrossov

During the day (February 27, 1943) approximately twenty bunker positions were taken. The enemy counterattacked in the evening and Matrossov's unit was surrounded. Not until dawn was the encirclement broken. During an attack on the main bunker in the enemy position the entire battalion could not advance because a bunker commanded the entire battlefield. Then-soldier Matrossov crawled up to the bunker and threw

his body in front of an embrasure. He was fatally wounded in the process. The bunker was then taken and the attack in this sector could be continued.

The abbreviation OKB (ОКБ) means Development Design Bureau or Special (independent) Design Bureau. Henceforth the customary abbreviation OKB will be used. The design bureaus were created in various branches of industry. There was thus an OKB-1 in various branches of industry.

By 1941, the twenty-four years of Soviet power had also had other effects on the development of the Soviet Union. Lenin's idea "Communism is Soviet power plus electrification" and the development of heavy industry brought many positive achievements for the common people. In border regions national minorities were given a written language for the first time in their histories. Literacy was expedited. In the 1930s, people from the Central Asian republics received the title professor for the first time. Despite the Stalin's *dekulakization* and the purges that followed, a new society with different values arose under the leadership of the Communist Party.

A female instructor at a Soviet military academy formulated it thus: "He lay there drunk, the *muzhik* (farmer). But when the party called—the Motherland is in danger—then he stood up, picked up his rifle, and defended his homeland." The German military found this out in the early weeks of its invasion of the Soviet Union.

THE WAR BEGINS

A Film Excerpt

For the families in the west of the Soviet Union the start of the war came as a surprise. A day of retreat, ten days, a hundred days.

Political commissar Sintsov, soldier Solotaryov, and the female military doctor Tanya, who was ill, made their way through the forest.

"If there are no Germans there, we can go right in. We are people, are we not?"

They were suddenly depressed to discover that in their own country they had to send a scout into a house into which earlier, before the war, he and any other would have carried a sick woman in his arms at any time of day without hesitation.

"I don't believe there are swine in there," he said. "And if there are, we still have the rifle."

With the doctor in his arms, he reached the door of the house and kicked his foot against it. A terrified fifteen-year-old girl pulled back the bolt. She saw a tall, broad-shouldered man with a haggard face, who was holding a woman wrapped in a tent square in his arms. His large hands trembled from exhaustion. She immediately noticed that he wore the red star of a commissar on both sleeves. Behind him stood a smaller man with a torn leather jacket holding a rifle.

"Let me in," said the tall one in a voice used to being in command, "where can I lay her down?" He looked into her shocked eyes and continued in a friendly tone: "You can see that we are in trouble."

The girl opened the door, and Sintsov, with the doctor in his arms, stepped into the room; with a quick glance he looked it over. The room was decorated half country, half city: a Russian stove, against the wall a wide bench, a buffet, a table covered with washcloths, wall boards with paper tips.

"Is anyone else here?" he asked the girl, the doctor still in his arms. From behind him a voice said, "Of course, why not?"

Sintsov turned half and saw, standing in the door to the next room, the man Solotaryov had mentioned. He was not very young and rather stout; his hair hung down in disarray and on his puffy face was a thick, light-brown stubbly beard.

Biryukhov turned the wick a little shorter and placed his elbow on the table.

"Tell me something, comrade political leader: what is going on? There you sit before me, the Red worker and farmer's army, and I respect you for not having taken off your uniform, and so I ask you. What is going on and how long will it continue? Don't think you're the first I've asked this question. I have spoken to soldiers, including a lieutenant who lived here. He oversaw the cutting of wood, but he knew damned little. Then there was a general; he had led a division. It had been in our forest before it went to the front. He was a battle-tested general, one couldn't say anything against him. He and his men had fought their way back from the frontier, then he reassembled his division and went back to the front. I said to him: comrade general, you have definitely never dreamed that you would fall back to here, never believed it. You don't need to tell me, I know myself that you never considered it! But things turned out differently than you imagined! Tell me honestly what you think now. Will you leave here? Will the Germans come into my house?"

With these words Biryukhov raised his head and let his gaze wander slowly through the room, as if he was about to take leave of it. "And what

was his answer? I left that out! Tomorrow, he said, we will go back into battle, we will give the Germans a proper thrashing and first of all drive them out of Yelnya. And what should I say? The division really moved out, thrashed them, and drove them out of Yelnya. But what about now? From here the general advanced and took Yelnya, but yesterday the Germans were already behind us. And how far? It is said that a telephonist from Ugra called Znamenka yesterday and they answered her in German. And that is fifty *verst** further east!" (**verst,* is an obsolete unit of Russian measurement).

"That can't be," said Sinzow. "You'll see how it is! The general took Yelnya, but the Germans are already in Znamenka. Where is this general now? Tell me that!"

"Where, where?" Sinzow suddenly became angry. "He is fighting somewhere in the pocket, and we will do the same, if not so surprisingly . . . It is what it is, anyway we have come from Mogilev to Yelnya. We could have lay down our weapons somewhere but we didn't. And the others, are they about as bad as us?"

"May be that they are not worse, but the Germans have encircled you again!" Why did we let things get to this stage?

The images are from the film of the novel *The Living and the Dead* by Konstantin Simonov. This question is put more sharply in the film than in the novel.

AN ANSWER WITH SERIOUS CONSEQUENCES

The answer was very simple, the consequences were unpredictable; "Something like this must never happen again!"

Military intelligence gathering and espionage were stepped up even while the war was still going on. The apparatus of repression scrutinized friend and foe, searching for possible enemies of the people. Soldiers returning from German captivity and forced laborers automatically came under suspicion. Companies working in the military-industrial complex were separated. The government created so-called black areas and closed cities, which few were allowed to enter and which appeared on no maps.

These areas were supplied with highly qualified scientists and engineers. The names of outstanding personalities were not revealed to the public or to other countries. Photographing or filming of important sites was forbidden. Technological and scientific advances were therefore not transmitted to the entire Soviet economy, so that no one could determine the capabilities of the Soviet defense industry. This was one of the reasons for the demise of the Soviet Union. The military-scientific-industrial complex was a state within a state.

Sarov, renamed Arsamas-16, a secret city, is now Sarov again. The Sarov Monastery's Feast of the Assumption Cathedral at the beginning of the twentieth century.

KATYUSHA

How was it that a weapon received this woman's name?

At the beginning of 1939, the headquarters of the artillery gave the RNII (РНИИ – Reactive Scientific Research Institute) the task of developing rocket projectiles based on powder. Operational trials were completed in late 1940. The High Command of the Red Army accepted the weapon in May 1941, and on June 21, 1941, the last day of peace, the government decided to put the ROSF-132 high-explosive-fragmentation rocket—called M-13 and the launcher BM-13-16 into mass production. "BM" meant combat machine, and the "16" stood for 16 launch rails.

After the war started, the first rocket launcher battery was formed using the seven available prototypes. Under the command of Capt. Frolov, the battery went to the front with 3,000 projectiles. Ballistic tables for the launchers were worked out during breaks in movement. Maximum range was achieved with a maximum speed of 1,017 feet per second and a launch angle of 45 degrees.

The name "Katyusha" soon became known throughout the Red Army. To the Germans, the rockets, approaching in parallel, sounded like organ pipes and this resulted in the name "Stalin Organ."

The *Wehrmacht* High Command first identified the rocket launcher as a new weapon in November 1941, outside Leningrad. From then on the rocket launcher was a standard weapon in the Soviet Army arsenal.

July 14, 1941:
The sky is blue and clear. Flyorov looks at his wristwatch: fourteen thirty. He steps back up to the battery commander's telescope. Before him spreads the spider's web of railway tracks. Locomotives are shunting, trains are being formed, trains for the fascist advance to the east. Flyorov realizes what's at stake: the tracks lead to Smolensk and on to Moscow. As of today the fascists are halfway to the capital. They must be stopped. The high command must gain time to defend the metropolis and assemble troops for a counteroffensive.

Krivochapov, who is not so easily unsettled, rubs his fingertips together nervously. "Is everything accurately calculated? It is the first salvo and we must report on it."

Flyorov's eyes shine. "Everything is correct. I am moving the launchers into position."

The seven three-axle trucks emerge from the darkness of the forest. Flyorov's order is received by radio. The big engines roar and the rocket launchers are driven to a shallow depression running in front of the forest.

The frames supporting the launch rails, on which the rockets sit, stand motionless but filled with dynamic power. Engineers Popov and Shitov, together with the commanders, check the three firing mechanisms. Then Popov gives the order: "Everyone take cover!"

Now only the drivers and the commanders are at the launchers. They wait for the order to fire. Then it comes: "Battery salvo!"

The first action.

With an ear-shattering howl, more than a hundred rockets hiss toward the sky. Behind them they leave long glowing trails, like comets. Now it is eerily still. The quiet lasts a few seconds.

Fuel cars explode. Ammunition cars that have been hit multiply the din. Guns are smashed against the locomotives waiting under steam. In seconds the rail beds are engulfed in a sea of flames. Above them spreads a huge cloud of smoke, darkening the sky.

One day later:

The commanders of the firing platoons report ready to fire. The fascists' heavy tanks are already driving toward the ferries. Flyorov waits, he wants to send them to the bottom in the middle of the river. The transports begin to move. The captain has the telephone in his hand. With an emphatic voice he orders: "Battery – salvo!"

Silver arrows: one hundred and twelve rockets howl towards the enemy in a broad arc, descend at a steep angle, and strike the target. Ferries disintegrate, tanks and guns sink in the river. Ammunition explodes like fireworks before mingling with the water. Impenetrable swaths of smoke cover the river. The enemy bridgehead breaks up. The fascists flee in panicky fear, trying to swim to the safety of the bank. Red Army troops leap from trenches and attack, chasing the enemy across the river and bringing back hundreds of prisoners. Krivochapov interrogates a Wehrmacht major. When asked what he thought when the rockets struck, the pale officer replied: "I thought that a hundred guns had fired at us at once. I have never experienced anything like it. They didn't tell us that the Russians had such weapons."

The battery's official name was "Independent Experimental Battery of the Reactive Artillery of the High Command Reserve." The soldiers loved it. Various girl's names were proposed and then rejected.

The fat Vorobyev joked good-naturedly: "Makhnutin probably won't melt the ice, but I have a good name. The SB bomber aircraft is called Katyusha by the airmen. When our rockets strike the target it is like an aerial bombardment. We should call our weapon that."

"I don't quite understand what you just said," murmured Makhnutin, "but I agree with Katyusha."

From *The Katyusha's First Salvo* by Heinz Bergmann.

By the end of the war there were more than 500 rocket launcher battalions serving in the Red Army.

LAUNCH SITE HEIDELAGER

After the bombing of Peenemünde on August 18, 1943, the decision was made to set up an alternate launch site in Poland. SS training Camp Heidelager was located west of the town of Rzeszow, enclosed by the Pilica, Vistula, Visloka, and San Rivers. At its center was the village of Blizna.

From mid-November 1943, until the area was occupied by Soviet forces, 139 A-4 rockets were launched there, and this was documented by scouts of the Polish Home Army. Maximum firing range was 155 miles.

The designers at Peenemünde concluded that approximately sixty percent of the rockets that were fired broke up in the air and failed to reach the target. Other problems caused the rockets to catch fire, and sometimes the guidance system failed. Other rockets never got off the launch pad or lifted off and then fell tail-first to the ground. Polish farmers used pieces of metal from them to cover their buildings. Testing of the A-4 was carried out over populated areas. The number of Poles killed during these trials was approximately 150.

The launch of an A-4 rocket.

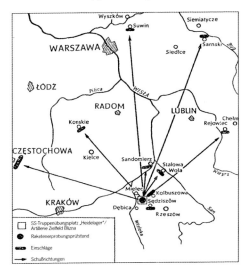
SS firing range at Blizna.

In the spring the firing concentrated on the town of Sarnaki. Members of the Polish Home Army collected and hid every fragment of rocket they could find. One particularly lucky incident took place on April 24, 1944. A farmer witnessed a rocket come down, and the complete guidance element was recovered. On May 20, a complete A-4 came down in the swampy bank of the Bug River. After the Polish Home Army had concealed the rocket with reeds and camouflage netting, days later horses were used to haul it from the bank of the Bug to a nearby village, where it was disassembled and photographed. Significant parts were delivered to scientists at the Polytechnic Institute in Warsaw.

The most important parts of the rocket were loaded into an aircraft. From a secret Home Army airfield, on July 26, 1944, the freight reached the town of Brindisi in southern Italy. The British also received the remains of a rocket that had crashed in Sweden. Churchill wanted clarity; he, therefore, waited no longer and turned directly to Stalin.

Stalin was in no hurry to grant the British access to the launch site. Both the British side and the Russian had—if at all—only vague ideas about the A-4 rocket. The launch site was unknown to the Soviet side.

Stalin's instructions were immediately passed to the general staff. Army intelligence subsequently received orders to give special attention to investigating activities in the Debica area. In July 1944, the front was just thirty miles from the launch facility. At the same time, Stalin ordered Shakhurin,

head of the People's Commissariat of the Aviation Industry, to prepare a group of Soviet rocket experts to study everything discovered at the launch site before the arrival of the British.

After the area was liberated, the first military intelligence expedition was sent, under General Serov. Members of the group included J.A. Pobedodosvev and M.K. Tikhonravov. The British group arrived a week later. An intelligence officer had a detailed map of the launch site with the coordinates of rocket launches and their crashes. The following work was very successful.

After the parts that were found were packed up and loaded, they were sent to the NII-1. The term A-4

Personal and Top Secret Message from Sir Winston Churchill to Marshall Stalin, July 13, 1944:

"There is reliable information to the effect that for a substantial period of time the Germans have been conducting missile tests from an experimental station in Debica, Poland. According to our information this projectile has an explosive charge weighing approximately 12,000 pounds, and the effectiveness of our countermeasures depends to a significant degree on how much we can find out about this weapon before it can be used against us. Debica is located on the route of your victoriously advancing troops and it is completely possible that you will seize this site in the next few weeks.

"Although the Germans will most certainly destroy or haul off as much of their equipment at Debica as possible, you will probably be able to obtain a great deal of information when this area is in Russian hands. In particular, we hope to find out how the missile is launched, because this will enable us to determine the missile launch sites.

Winston Churchill.

"I would therefore be grateful to you, Marshall Stalin, if you could give the order for the devices and equipment to be preserved when your troops have taken this area and subsequently give us the opportunity for our experts to inspect this experimental station."

Exchange of Letters Between the Chairman of the Council of Ministers of the USSR, and the President of the USA, and the Prime Minister of Great Britain During the Great Patriotic War, 1941–1945, second edition, vol.1, Moscow, Politisdat, 1986.

was not yet known. All of the parts were stored in the institute's ballroom. Only selected personnel were allowed to enter, including Prof. Bolkhovitinov, A.M. Isayev, H.A. Pilyugin, W.P. Mishin, and B.E. Chertok.

Isayev could not restrain himself and crawled through the nozzle into the combustion chamber.

"What is that, Viktor Fedorovich?"

"It is something that ought not to exist!" was the answer.

Soviet rocket engines of that time produced 3,300 pounds of thrust, but the A-4's engine was capable of producing 55,155 pounds. More parts arrived from Heidelager. The group headed by Prof. Bolkhovitinov was given the task of reconstructing the structure of the entire rocket, its control system, and its important technical parameters.

In assessing the capabilities of the Soviet R-1 and R-2 rockets, an examination of the reliability of the V-2 rockets is of interest. *SS-Gruppenführer* Hans Kammler published the following report in December 1944: "The information reveals that of the 626 rockets, 131 failed to launch (20.95%). Information about failures during the descent phase was not gathered, which meant that only about 400 rockets (62–63%) must have reached their targets. The rocket was thus not very reliable and the transport of just over a ton of explosives was not very effective."

Ruth Kraft, one of Wehrner von Braun's coworkers, told the author that at Peenemünde only the term A-4—*Aggregat 4*—was known and used. The designation V-2—for *Vergeltungswaffe-2* (Revenge Weapon 2)—did not become common parlance until Nazi propaganda minister Joseph Goebbels used it in October 1944.

Rockets delivered to the military	626	
Faulty guidance systems prior to launch, returned to the factory	87	12.3%
Launched	495	
Percentage of failed attempts		44%
– Defective guidance systems		41%
– Defective propulsion		13%
– Fire in aft section of rocket		13%
– Explosion at launch		2.9%

CHERTOK IN BERLIN AND PEENEMÜNDE

Toward the end of the Second World War, gathering trophies of the defeated enemy acquired ever greater importance. On April 16–17, a group was formed under Gen. Popov which was given the special powers by the State Defense Committee to inspect, examine, and if necessary, seize samples and material of German radar technology and equipment technology.

The Petrov group was divided into teams of three. One of these groups, consisting of Smirnov, Chistyakov, and Chertok, had a special mission: "Study German aircraft, autopilots, aircraft armament, and aircraft radar, navigation and radio technology."

In the Moscow military command these "civilians" were fully outfitted as officers, with corresponding ranks and pistols with ammunition.

On the morning of April 23, the group took off from the central airfield at Frunze, flew over the western areas of the Soviet Union, over Poland, and landed at a forward airfield near Strausberg just behind the front.

After Germany's surrender, many Soviet engineers, rocket experts from the different fields of technology, were active in the Soviet zone—among them W.F. Bolkhovitinov, A.M. Isayev, B.E. Chertok, W.I. Kuznetsov, W.P. Barmin, W.P. Mishin, N.A. Pilyugin, S.P. Korolev, and W.P. Glushko.

At Peenemünde they saw not only the V-2, but also a series of smaller missiles: *Rheintochter*, *Rheinbote*, *Wasserfall*, and *Taifun*.

Boris Chertok reports:

"We looked at one another and spoke with the pilots. Finally Gen. Petrov appeared and assigned us to three villas in Straussberg. The officer's mess astonished us: unusual cleanliness, glaring lights, waitresses in neat skirts and bonnets. Where were we? It was unlikely that deadly fighting was raging just a few kilometers away.

"In then morning we were assembled by the general for instructions and to come up with a work plan. Our first task was to carry out a comprehensive investigation of the delivers, the German aviation research institute in Adlershof. We were to kook for 'devices' that had been thought up by God knows whom. Our instructions read: examine German factories and laboratories, Intellectual achievements are not important, instead the most important thing is to take inventory of the types and numbers of workbenches, technological and production equipment and the measuring devices. As far as the documentation and specialists went, this was already a question of our knowledge and our initiative.

"For two days we intensively studied the maps and travel routes. We gathered the addresses of the companies and facilities in the greater Berlin area we were interested in. Finally, on April 28, we set off through the streets of Berlin for Adlershof. We began with the detailed examination of the buildings in Adlershof. It had not been the scene of any particularly fierce fighting and all the buildings remained. At the entrance behind a large bicycle rack, key after key after key hung on a large rack—with German thoroughness each with a number. All in order.

Boris Evseyevich Chertok.

"We inspected the DVL—administration buildings, archive, files, personnel documents in a safe. A sergeant with a soldier knew where everything was.

"The soldier placed a large crowbar on the door of the safe. The sergeant struck the crowbar with an accurate and powerful blow from a heavy sledgehammer. As usual, the safe was open after the first blow. It was full of reports with red stripes: 'Secret' or 'Top Secret.' We took the reports with us. Reports about every possible kind of experiment.

"DVL—that is the equivalent of our TsAGI, LII and NII of the air forces combined. We had no time or physical opportunity to read and study. The general gave the order for everything to be labeled, packed in boxes, and flown to Moscow. But where to get the boxes and how many? It turned out that the air force battalion was capable of doing anything, had everything we needed, and organized everything! And there was no need to label the boxes . . .

"I must admit that I sinned. I took one report and I still have it today. It is a work by Dr. Magnus about damped and gimbal-mounted gyroscopes as measuring and sensor units for angular velocity.

"Laboratory buildings, an aircraft navigation laboratory filled with test benches for testing onboard equipment, a photochemical laboratory, a laboratory for testing the strength and fatigability, vibration stands, a test bench for bomb and gun sights, and a device for balancing gas pedals. And what outstanding drafting and design facilities! The German design workplaces aroused feelings of envy in us. They also had good swiveling chairs, comfortable work tables with plenty of drawers filled with small items, and everything in its place. Oh this German love of details and accuracy, which represented a special class in working culture.

"The most important and indispensible items for every laboratory were the four-channel oscilloscopes from Siemens. We found a variety of types there: two-, four- and six-channel ones. Without these devices the investigation of fast-pace dynamic processes would be impossible. This was a new stage in measuring technology and engineering research. In the NII-1 institute in Moscow we had one six-channel oscilloscope for the entire institute. No, already we no longer felt hate or the desire for revenge which at first had boiled in all of us. We already regretted that we had been forced to break open such solid laboratories and entrust the packing of such invaluable precision equipment into boxes to energetic but not especially careful soldiers.

"But faster, faster—all of Berlin was waiting for us! I climbed over the body of a very young German soldier with a panzerfaust which had not yet been removed and with my column from the air force battalion prepared to open the next safe.

"It was a fantastic electro-technical measuring laboratory. Inside we discovered a large number of (to us) unique devices made by the world-famous German companies Siemens, Siemens & Halske, Rohde & Schwarz, the Dutch company Philips, Hartmann & Braun, and Lorenz. We also found: photographic enlargement equipment, photo projectors, cinema projectors, chemicals, large stationary cameras, cinetheodolites, photo theodolites, as well as optical devices whose purpose was unknown to us.

"Another unique building was dedicated to electrophysics. Electric frequency meters for low and high frequencies, wavemeters, precision sound level meters, octane filters, harmonic analyzers, harmonic distortion meters, engine generators and transformers for various voltages, even cathode ray oscillographs (one would now say electronic oscillographs). The richest building complex was the one with the radio and sound measurement equipment.

"On the boxes we wrote the addresses of our companies: 'name/I am so and so.'

"I had not discovered such a large quantity of secret and top secret reports for a long time as those I send from Adlershof to Moscow. They were distributed to the LII (Aviation Research Institute) and the TsAGI and to the NISO and other facilities of the aviation industry. NII-1 received about a dozen boxes of the measuring technology we sent. My immediate superiors were obviously impressed.

Boris Chertok at the 14th Day of Spaceflight 1998 in Neubrandenburg. He spoke very frankly: "And in 1945, we did the same thing as the Americans (with respect to German rocket experts). We took that we could catch."

"On the evening of May 2, we made our way to the Askania Company. We had become aware of Askania while still in Moscow. At the DVL we also found clues about this company's diverse activities. And then we were on our way there. The military commander of the district had been appointed on May 1st, however he already had a functioning mayor. He wore a red armband with the title 'Bürgermeister.' After listening to us, he took out a map of the city and explained to us clearly where we should look. He was amazed and said: 'This facility is very small. It is only a branch office of Askania.'

"We nevertheless found the operation very interesting. It manufactured gyroscopic horizons, directional gyros for the V-1. And an American-type magnetic directional gyro had just been put into production. 'Isn't this an exact copy of the Sperry?' we asked. 'Yes, that's correct, we study the American technology from shot-down aircraft. We must admit that the Americans are far ahead of us in remote devices', was the answer. As directed, we checked and inventoried all of the workbenches. We were particularly pleased with the precision drilling machines with a large number of operating speeds from 500 to 15,000 rpm and with flexible regulation.

"On May 3, we received information that we could not delay our inspection of the western zone of Berlin, because in May this part of the city was supposed to be handed over to our allies, after which we would not be able to enter or it would become pointless. At that time there was not a single allied soldier there.

"We continued our inspection of the Askania Company in the days that followed. It was a company with a broadly diversified production profile and a competitor of Siemens. In Mariendorf we found a large factory and a design bureau. There we ultimately discovered a complete and undamaged guidance system for the V-2 and a very similar one which had been designed for aircraft autopilots. Completed autopilots were on the test benches, ready for delivery.

"We were astonished to find a department with periscopes for submarines and associated range-measuring equipment as well as bombsights and control systems for anti-aircraft artillery. Special rooms had been set up for training crews, in which the operation of all aircraft equipment associated with instrument flight was simulated.

"There was a rather large department engaged in the production of purely optical equipment. The grinding lathes for optical lenses and the finished products sat side by side. We found lenses of different diameters up to twenty inches!

"There were very well-equipped experimental laboratories. We also found: barometric chambers, thermo-barometric chambers, vibration stands, rain simulators. And everything equipped with universal and specialized measuring equipment and—our dream—multichannel oscillographs made by Siemens.

"On May 8, we inspected another Askania factory in Friedenau. There we met with the firm's technical director. He showed us a sketch (there was no finished drawing) of a polarization relay for the actuator of a V-2. He told us that all departments of his firm had the most modern collection of measuring technology and workbenches in Europe. In particular he extolled the coordinate drilling machines and optical benches.

"As we continued our work, conflicts developed between the various Soviet authorities. The first took place the same day, May 8, when we drove into the Askania Gyroscopic Equipment facility. At the entrance stood two men armed with submachine-guns: 'Comrade officers, we cannot let you in.' After an exchange of words one of the men disappeared and came back with a lieutenant colonel. He was a 'union officer' like us. Sinoviy Moiseevich Zezior introduced himself and apologized for having to prevent a representative of the aircraft industry from entering, because the entire facility was being handed over to the shipbuilding industry.

"We quickly became friends with him and he remained with us for many years as an advisor in the development of gyroscopic equipment for missiles. We were given permission to examine the gyro platforms which had already entered production. According to company experts, two years earlier they had received an order from Peenemünde for guided projectiles. Exactly what kind only the senior company officials, who had fled to the west, knew for sure. Viktor Ivanovich, later chief designer of gyroscopic equipment for missiles and space equipment, a future member of the academy, twice Hero of Socialist Labor, but at that time a very tall and thin colonel in a field blouse that was much too short for him, told us about the gyro platform and

about the computers there for lateral and longitudinal acceleration. Kuznetsov declared: 'It is a very complete product. We are making passable equipment for ships, but for rockets and on this scale?!'"

May 9, 1945:
"Our inspection of the Telefunken plant in Zehlendorf was very interesting. The plant initially produced vacuum tubes and was later almost completely converted to the production of radar equipment. In contrast to most other facilities, here we found almost the entire work force including the chief engineer Wilki and his closest coworkers. Chistyakov and I already spoke relatively good German at that time, and we needed no interpreter.

"Wilki and the head of production showed us the factory and laboratories. Wilki was head of research in the field of centimeter-waves, and American and British radar devices installed in aircraft were studied closely in the laboratory located outside the factory grounds. The same was true of target-finding equipment used by bombers and reconnaissance aircraft. The German specialists estimated that the British and Americans were furthest advanced in signals intelligence. This was especially true of the war against the U-boats. Their aircraft were capable of detecting periscopes from dozens of kilometers away. The Germans therefore worked on equipment to inform submarine crews when they had been detected by aircraft radar. The factory built aircraft search radars based on British and American experience. The department for production of radio location devices was well equipped with electronic measuring equipment. The factory was relatively new, having been completed in 1939. Including Ostarbeiter, six- to seven-thousand people were employed there, including three-thousand engineers and technicians. There were no problems with the supply of materials.

"The large screens for the radars and receivers were made by the Lorenz and Blaupunkt companies.

"Haven't you investigated Soviet radar systems?" we asked.

"'According to our military, no such equipment has been found in any of your aircraft. And none that would have been of interest to us was in the goods captured by our troops during their advance. We came to the conclusion that the Russians had shielded this technology so well that none of it fell into the hands of our military.'

"I think that he used the term 'shielded' out of courtesy. In fact they had long since guessed that virtually no radar systems and no radio-location technology was installed in our aircraft during the Second World War.

"We questioned the German specialists further about the other companies and their research work. Like all radio and electronics specialists they were well informed about the secret companies and their research and told us that the air defense's radio location technology was made mainly by Telefunken and Lorenz and the remote guidance technology by Askania and Siemens. In the past year and a half many leaders with their personnel and laboratories had been moved to Thuringia and Westphalia. They knew that the 'secret weapons' were being developed at Peenemünde. None of them had been there as it was top secret. Another department of Telefunken built ground-based location equipment and rocket radio-control systems.

"The tube production departments were very well equipped. They produced Magnetron tubes (magnetic field tubes) with a pulse power of up to 100 KW!

"In front of the gatekeeper's lodge we were surrounded by a throng of people, mainly women.

"'Sirs, we would like to know what will become of us. Are you going to ship us to Siberia?'

"'You will neither be imprisoned nor made prisoners. As far as active Nazis, your mayor will deal with that."

"'No, you didn't understand us. When will you give us work? And who will pay us? Or do you not need the equipment we can build?'

"Honestly these were difficult questions, and this just five days after the fall of Berlin."

May 10, 1945

"With difficulty we made our way to the Lorenz Company in Tempelhof.

"The plant itself had been occupied by 'union officers' from the Moscow radio works before our arrival. We spoke with the German experts for about two hours. They showed us radars with wavelengths of three and nine centimeters (1.18 and 3.54 inches). It was interesting that the laboratory, which had specialized in the development of radio receivers, had been hastily re-profiled for the development of equipment with large cathode ray tubes for use with radars.

"The facility manufactured ground radio stations with large rotating antennas for guiding aircraft on approach to land. We convinced ourselves that these radars could also be used to control aircraft in combat within direct visual range. And we were amazed by the large number of stations with a panoramic view and a large radar screen, which made it possible to recognize enemy aircraft and tell them from one's own. The Germans declared that they had made several hundred such stations. This was difficult to verify, when one considers the very complicated construction and the great amount of work required to produce the systems. The Freya radar had been developed in 1938. It was able to detect aircraft at a distance of seventy-five miles.

Würzburg-Riese radar for the detection of aircraft formations. Janick/Neumann Collection

"The Würzburg radar, with a spherical antenna, had been developed as a fire-control system for the anti-aircraft artillery. The more powerful Würzburg Riese was used to guide night-fighters to their targets. When the war began, the entire German radar technology was oriented on the decimeter waveband. The German engineers told us: 'Our battle with the British was not just the war on the ground and in the air but also in the laboratories. They achieved great success in 1942, by the clever switch to the centimeter waveband. At that time we had no equivalent tube technology.'

"After lengthy talks with the German radio specialists, we left the Lorenz radio works and met with our colonel. The colonel proved to be just as much a 'union officer' as we were. Alexander Ivanovich Shokin was the representative of the Radar Council of the State Committee for Defense. At that time I had no idea that I was speaking with the future deputy of the minister responsible for the radio-electronic industry, who later rose to become Minister of the Electronics Industry.

"Back then in Berlin he observed bitterly that, despite considerable scientific success, our radio-electronic and electronic industry undoubtedly had to be assessed as weakly developed compared with what we saw here. As in all of our previous visits to German businesses and laboratories, we were impressed by the wealth of metrological equipment compared to our limited assets. It was both universal and specialized. There were valve voltmeters, oscillographs, sound generators, banks of every possible kind of filter, standardized amplifiers, wavemeters, frequency meters, and many others, all of the highest quality. Some devices, which before the war we regarded as treasures, were in fact in every institute. None of our institutes, factories, or laboratories could even imagine such an abundance.

The laboratory war was not just a war of pure intelligence. Each intellectual had to have access to the most modern research instruments—this guaranteed the accelerated development of the equipment engineering industry.

"At that time we collected important-looking literature and, on which I insisted, captured measurement equipment and sent it to Moscow.

"By mid-May, our troika was able to draw a more-or-less-clear picture of the equipment and

Würzburg-Riese radar. Cosford Luftwaffe Museum, private collection A. Kopsch

radio industry in greater Berlin. There were more than thirty businesses on the list.

"We were not just interested in individual businesses, rather in the entire organization and structure of the equipment and radar industry."

Flight to Peenemünde on June 1, 1945

"The staff which was to investigate Peenemünde was led by Major General Andrei Iliaronovich Sokolov. During the war he had been the deputy commander of the guards mortar units.

"Almost no German specialists remained on the island of Usedom. The group around General Sokolov chose several less-informed specialists from among the local inhabitants. With their help and the intuition of the Soviet engineers they worked up a description of Peenemünde's function—how it was, but no longer is. The Allied air forces had damaged almost all the buildings. There was, however, no destruction down to the foundations. The scale of the firing positions surpassed all of our expectations.

"In addition to the firing positions, there were well-preserved bunkers from which the testing of engines and rockets was controlled and observed. All of the installations, which covered some sixty acres, were connected by excellent roads. Dozens of miles of high-tension, measuring and signaling cable had been laid in the cable ducts. The Germans had failed to dismantle it. All of the equipment had been dismantled, down to the last screw, including in the big factory, which was almost undamaged. Everything had been transported away and a special detachment of the SS had made unusable everything that could not be evacuated before the arrival in Peenemünde of Marshall Rokossovsky.

"Peenemünde brought another secret to light.

"A soldier found a thin booklet bearing the words 'Top Secret'. It was the expertise produced by a collective about a project for a rocket-powered bomber aircraft.

"Isayev told us about it. 'Our BI rocket aircraft produced 3,300 pounds of thrust, this one has 220,000 pounds of compact thrust. With this devilish engine the aircraft can reach an altitude of 180 to 240 miles. Then it descends at supersonic speed but does not enter the atmosphere, instead striking it like a flat stone does the surface of the water at a minimal angle. It meets the atmosphere, shoots upward, and flies on! Skipping flight, and it does this two or three times. The aircraft glides through the atmosphere in this way and then, after it has crossed the ocean, it comes down in a dive to attack New York. A tremendous idea . . .' The discovered booklet was now declared secret for a second time and hidden under the shirt of Isayev's trusted coworker. This booklet spoke of ranges that were necessary to attack New York.

"The unique find was taken to Moscow by a special flight and passed directly to our patron, General Bolkhovitinov. He evaluated the for-that-time sensational find with Engineer Hollender, who had good command of the German language.

"This work had appeared in Germany in 1944. Its authors were the Austrian rocket researcher E. Sänger, who had become known before the war, and an I. Bredt, who was unknown to us. E. Sänger had become known through his book *Rocket Flight Engineering*, which was published in 1933. It had been translated and appeared in the Soviet Union. Engineer Sänger, who was just 25 years old, was involved in rocket research. He was one of the first researchers to seriously explore gas dynamic and thermodynamic processes in rocket engines.

"One can imagine what Bolkhovitinov and the other staff of the NII-1 felt when they read the top secret report. One hundred copies had been printed and issued to the officers of the Wehrmacht High Command, the aviation ministry, and all institutes and organizations working in the field of military aviation, as well as to all specialists and senior staff involved with rocket technology, including the armaments department under

Dornberger, who was then head of the Army Research Institute in Peenemünde. The report bore the title 'A Long-range Bomber with a Rocket Engine.' In the work the technical possibilities of a very large controllable winged rocket were thoroughly analyzed. The authors had convincingly shown and calculated, and had produced nomograms and diagrams which showed that with the rocket propulsion system they were proposing, producing 220,400 pounds of thrust, flight at altitudes of 30 to 185 miles was possible at speeds of 12,400 to 18,600 miles per hour and a range of 12,400 to 24,850 miles. The physical-chemical processes of propellant combustion under high pressure and high temperatures, the energetic properties of the propellant, including that of an emulsified light metal-hydrocarbon combination, were extensively researched. A closed steam-powered direct current device was developed, which cooled the combustion chamber and powered the turbopump system.

"The problem of the aerodynamics of aircraft flying at Mach 10 to 20 proved to be something new for our aerodynamicists. The report also described launch systems and the dynamics of launch and landing. The questions of aiming bombs dropped at the high speeds of which this aircraft was capable was examined in great detail, obviously to attract the interest of the military. It is interesting that as early as the beginning of the 1940s, Sänger was able to show that additional systems were required to launch an aircraft into the cosmos. The authors proposed a catapult launch system from a horizontal position. Taking off from a horizontal launch trolley, the aircraft was to achieve an initial speed of Mach 1.

"In their calculations and diagrams the authors commented on the launch as follows: 'Takeoff is achieved with the help of a powerful rocket, which is restricted to the ground and operates for approximately eleven seconds. After reaching a takeoff speed of 1,640 feet per second, the aircraft lifts off from the earth and with the engine at full power reaches an altitude of 31 to 93 miles on an initial flight path 30% against the horizon, which then however becomes flatter. The climb duration is 4 to 8 minutes. As a rule, all fuel is consumed during this period. At the end of the flight path's ascending branch the rocket engine finishes its work and the aircraft continues its flight with the aid of the saved kinetic and potential energy on a wavelike flight path with diminishing amplitude. The aircraft drops its bombs at a predetermined time. The aircraft follows a large arc and returns to its airfield or other landing place, while the bombs continue in the old direction and reach their target. Such an attack tactic is completely independent of time of day and weather over the target area, and the enemy is powerless against such an attack. The assignment given us has never been accomplished by anyone anywhere. The goal is to fire at and bombard a target at a range of 600 to 12,000 miles. A unit of 100 rocket bombers would be capable of completely destroying an area the size of the world's capitals and their suburbs located at a chosen point on the earth's surface, in the course of several days.'

"The bomber's gross weight is 220,400 pounds, of which 22,000 pounds is taken up by the bomb load. Landing weight is 22,000 pounds. By reducing the range of the flights, bomb load can be increased to 66,000 pounds.

"Further development of the rocket bomber is to be divided into twelve stages. The main time is reserved for static testing of the engine, testing the combined effect of engine and aircraft, testing the takeoff system, and finally all stages of flight testing.

"Sänger and Bredt's work was translated in 1945, and in 1946, it appeared under the title 'Survey of Captured Technology edited by Maj. Gen. of the Air Force Engineering Services V.F. Bolkhovitinov' and was widely distributed by the Military Publishing Company of the Defense Ministry of the USSR.

"This project was at least twenty-five years ahead of its time."

All of the members of the future Council of Chief Designers were on an official trip to Germany to study German rocket technology.

THE VICTORS: AN INITIAL BALANCE

Peace finally returned to Europe on May 9, 1945. People breathed a sigh of relief—no more death, no more destruction. But victory had come at a terrible cost.

Losses USSR

- 41% of all losses of all the peoples involved in the war
- 27 million dead
- 25 million homeless
- Completely or partially destroyed: 1,700 cities, 70,000 villages, and more than 6 million buildings
- Also destroyed were 32,000 industrial districts, 40,000 miles of railroad tracks, 98,000 collective farms, 1,876 state farms, 2,890 tractor stations

Agricultural losses were immeasurable. During the war years the Soviet Union lost about thirty percent of its national wealth.

Losses USA

- 0.4% of all losses of all the peoples involved in the war, equivalent to 1.267 billion dollars.
- 300,000 dead

But the peace was soon placed in question by a new military threat. The term "Iron Curtain" entered the public consciousness.

Winston Churchill had coined the expression for the sealing off of the eastern block against the west. He first used the expression in a telegram to US President Harry S. Truman on May 12, 1945, a few days after the unconditional surrender of the *Wehrmacht*. Speaking to a large audience, on March 5, 1946, he declared in Fulton, Missouri:

"From Stettin in the Baltic to Trieste in the Adriatic an Iron Curtain has descended across the Continent. Behind that line lie all the capitals of the ancient states of Central and Eastern Europe. Warsaw, Berlin, Prague, Vienna, Budapest, Belgrade, Bucharest, and Sofia; all these famous cities and the populations around them lie in what I must call the Soviet sphere, and all are subject, in one form or another, not only to Soviet influence but to a very high and in some cases increasing measure of control from Moscow."

The Sinews of Peace

Not least because of its wording, this speech is considered to be "the clarion call for the Cold War."

The dropping of the atomic bombs on Hiroshima on August 6, and Nagasaki on August 9, ended the Second World War with a new weapon. From the history of the warfare we can draw the following conclusion: when the current war ends, the next begins. The American president left no doubt as to whom the next enemy would be.

GERMANY'S ECONOMIC AND SCIENTIFIC SITUATION IN 1945

Modern investigations lead to the conclusion that in 1939, Germany had a ten-year lead in rocket research over the rest of the world. Despite the war, further developments and projects were undertaken. As a result of the concentration of rocket development at Peenemünde, by March 1945, Germany's lead in rocket development had grown to twenty years.

In his memoirs, the control expert Boris Chertok, who later became Korolev's deputy, impressively described the truth of the above statements:

"All of the businesses were technologically and economically superior to our domestic operations. Most interesting were the laboratories of the companies Askania, Telefunken, Lorenz, Siemens, AEG, Blaupunkt, and Radio-Loewe. A new discovery for us was that there was a List company in Germany, which had grown very quickly and was engaged in the development and production of multi-contact plug-in connectors. Hundreds of thousands of these were manufactured for the entire aircraft and rocket technology fields. Technologically, these were very simple in design but completely new products, for with the complexity of the electric switches in aircraft systems, during the repair and testing of individual sections and for rapid assembly, it was necessary to be able to reliably connect and disconnect the electrical switches.

"After the war we even adopted the word '*Stecker*' (plug) from the German. There are many examples in history of the victor taking much from the defeated.

"It was not until after the war that we realized what great technical roles such seemingly simple components like plug-in connectors played in aircraft and rocket technology. The Germans spent years developing the technology of reliable connections and introduced the standardized 'List plug' with two

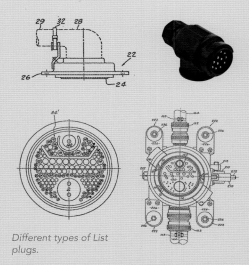

Different types of List plugs.

or three contacts. It took us three years to reproduce equally reliable connections. These connectors caused us more than a few problems for several when they were introduced into rocket technology.

"The Soviet telecommunications equipment used by East Germany's National People's Army received a wide variety of these plug-in connectors until into the 1980s.

"The German companies tackled a multitude of technical problems on their own initiative with no direction 'from above'. They did not have to wait for decisions by the state planning commission or people's commissariat, without which none of our companies could produce any sort of product. And in this lay the strength of the equipment and radio-technical industry. The electrical measuring equipment industry, as well as the equipment-building and radio industries, developed very positively until the start of the war. They dominated the European market and successfully competed with products from the USA. The companies Hartmann & Braun, Telefunken, Anschütz, Siemens, Lorenz, AEG, Rohde & Schwarz, Askania, and Carl-Zeiss were known around the world long before the Second World War.

"This created a sound technological base, whereas in our people's economy such branches were not present on the required scale when the war began.

"Our general electric equipment industry, the aircraft industry, and finally the shipbuilding industry were developed solely in several operations in Moscow and Leningrad (the companies Elektropribor, Teplopribo, and Svetlana in Leningrad, and Aviapribor, the Lepse Works, Elektrozavod, and Manometer in Moscow).

"It is noteworthy that when, after the war, we began reproducing the V-2 rocket technology and developing our new missiles, we had to convince ourselves that a device known for a long time—like an electric multi-contact relay—was only made in our country by a single facility in Leningrad, the Krasnaya Sarya factory. In Germany the Telefunken company alone had three such operations and Siemens at least two. This was one of the reasons why armaments production in Germany rose steadily until the summer of 1944, despite the uninterrupted bombing of German cities by the Allies."

That was a general economic comparison; but what about rocketry development?

Boris Chertok further wrote:

"The inspection of Peenemünde organized from May to June 1945, showed that the actual extent of the work on rocket technology far exceeded what we had imagined. We Soviet specialists had to grapple with the full extent of the German work in the field of rocket technology. It was no less important to find out about the complete development history of the rockets and become familiar with the working methods of the German scientists and engineers who had solved the difficult problem of developing guided long-range ballistic missiles.

"In 1945, neither we nor the Americans nor the British were capable of building liquid-fueled rockets producing more than 3,300 pounds of thrust. And those that had been built were unreliable, were not put into production, and were not helpful in developing new types of weapon. At that time the Germans had successfully developed, tested, and put into production liquid-fuel engines with a thrust of 55,100 pounds and more, or more than eighteen times as much! As well, they were producing these engines on an industrial scale.

"Sänger's report, describing an aircraft project with an engine thrust of 220,400 pounds, also appeared at the end of the war!

"And then the automatic guidance system! It is one thing to show that it is possible, in principle and theory, to control such missiles and their engines during their flight over 180 miles, but it is another to solve this problem practically and develop the system into an operational weapon! The Second World War produced three scientific-technological achievements which have completely changed earlier concepts of strategy and tactics. They are:

- automatically guided missiles,
- radar, and,
- atomic weapons.

"The first two did not require the discovery of any sort of new physical laws.

"Atomic weapons are different. Their creation became possible by discovering new physical laws, through the penetration of the micro-cosmos, and the foundations of the structure of matter."

In conclusion, one must find that, economically, the Soviet Union lay broken. With millions of dead, millions of missing workers, and rocket experts had to be replaced immediately. The people needed places to live, food, clothing. How could this be accomplished given such devastation?

But the new military requirements demanded very different objectives.

At the end of the war—Trümmerfrauen (literally "rubble women").

THE RABE INSTITUTE AND OTHERS

At the end of June the Americans began leaving Thuringia. Behind them came Soviet troops.

From conversations in Berlin and Peenemünde, the rocket experts knew that the "Peenemünde people" had been evacuated to Central Germany, to the area of the southern Harz Mountains, centered on the towns of Nordhausen and Bleicherode. The group arrived in bomb-ravaged Nordhausen in July 1945. Its most important task was the gathering of information. Answering a call by the mayor, many people who had worked in the "Mittelwerk" came forward. Lt. Schmargun, who had been a POW,

offered his help and proved very valuable. He led Isayev and Chertok into the "Dora" death camp.

In a dark corner, hidden under covers and rags, was a spherical object. After removing the rest of the material, those present discovered a gyro-stabilization platform, which mysteriously had not been installed in a V-2. For the Russian rocket experts it was a sensational surprise.

The group went to the Mittelwerk. At the entrance they were met by Engineer Rosenplenter, who came from Peenemünde. He told them that the Peenemünde people had been evacuated to

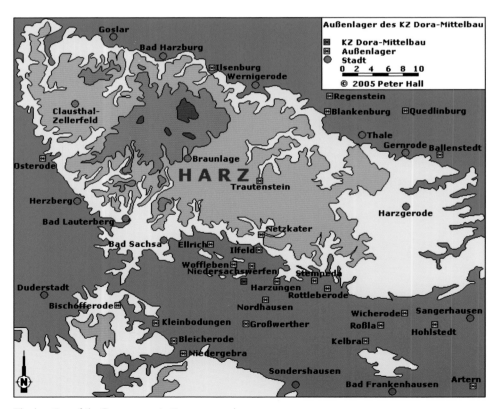

The location of the Dora concentration camp and its satellite camps (Außenlager) in the Harz Region of Germany.

Nordhausen and that von Braun and Dornberger were staying in Bleicherode. Inspection of the Mittelwerk lasted a total of two days. The result of the inspection was sobering. The methods of suppressing the inmates could not be escaped. After the Americans withdrew, the usual work benches and normal equipment were present in all areas. There were sufficient parts for ten to twenty rockets.

The camp inmates had sabotaged rocket production in such a way that their work was difficult to detect. Up to twenty percent of the rockets were found to be defective during final inspection at the Mittelwerk. Rockets with defects were transported to Factory 3 in Kleinbodungen near Bleicherode. The Soviets deduced that testing equipment must still be present there.

From Germans at the Mittelwerk, the two rocket experts learned that there was an engine test stand in the town of Lehesten near Saalfeld. All rocket engines installed in the Mittelwerk came from there. Further clues led Isayev and Chertok to

get into their automobile and drive to Bleicherode on the morning of July 18, 1945.

The local military commander assigned them a villa: the Villa Frank. Wehrner von Braun had lived there for a short time. The villa was occupied and served as a headquarters and officer's club.

By now the group consisted of twelve Germans, and Lt. Col.Isayev and Maj. Chertok as leaders. The Germans declared that the number of rocket experts would grow quickly and that the operation had to be given a name. A short "Soviet-German" consultation on July 25, 1945, resulted in the name Rabe Institute. Rabe was not a reference to the bird (raven), rather it was an abbreviation for "*Raketenbau*" (rocket construction), "Bau von Raketen" (building of rockets), "*Raketenbau Bleicherode*" (rocket construction Bleicherode), or "*Raktenbau und Entwicklung*" (rocket construction and development).

Creation of the institute was an unauthorized action, especially as the border of the American zone was only a short distance away.

The division commander was not responsible, and so Isayev and Chertok drove to Weimar to see the head of the Soviet Military Administration I.S. Kolesnichenko. The talks did not go well until an officer of the administration came up with the following idea: "The institute must be registered as a new scientific installation. Under the control of the military administration will gather scientists who we don't want to leave unemployed. As well, they will help us understand the secrets of the fascist secret weapons to find proof of war crimes.

Views of the Mittelwerke showing production of the V-2.

Villa Franke in August 2010.

Not only should we not oppose this initiative, we should support it with all the means at our disposal!"

I.S. Kolesnichenko assumed responsibility, and everything for the Rabe Institute was regulated: stamps, forms, telephones, work with German personnel, food, financial resources.

On his return, Maj. Chertok was named director of the institute. Other directors were Rosenplenter and Müller. The first task was to establish working conditions in the institute. The building, which formerly housed the mountain inspectorate, was adapted to meet requirements in one day.

The most important initial task was to reconstruct on paper all of the technology developed at Peenemünde. All parts not taken away by the Americans, those in the Salzburg factory and hidden or abandoned in other places, had to be gathered together. All of the rocket experts who had worked in Peenemünde were now scattered all over Germany, and they had to be found and seized. This included anyone who had worked in other companies and had any connection to rocket technology. Even the search in the American sector produced results, as the Americans were only interested in the leading figures of the rocket program. The number of rocket experts grew day by day.

The search for treasures was also successful. A complete Viktoria Honef radio guidance system was discovered 1,600 feet down a potash mine shaft.

In the period that followed, various rocket experts arrived to supplement the leadership of the Rabe Institute: Nikolai Alexeyevich Pilyugin (later chief designer of a scientific production association in the electronics industry), Leonid Alexandrovich Voskresenskiy (later Korolev's deputy for rocket testing), Vasiliy Pavlovich Mishin (later Korolev's first deputy and his successor after his death), Alexander Yakovlevich Bereznyak (developer of the BI-1 rocket-powered aircraft, later chief designer of winged rockets).

By the end of August, the institute was a large functioning operation. There were laboratories for gyroscopic devices, actuators, electric switches, ground-based control panels, and radio equipment. A design bureau was set up as was a photocopier

The former mountain inspectorate, which became the main building of the Rabe Institute, in 2010.

workshop. All that was missing was the technical documentation.

At the Kleinbodungen plant, in September, the assembly of rockets from power units and parts left behind began. The outer parts were sufficient for at least fifteen rockets, but control systems, engines, turbopumps and other equipment was not available.

From Moscow came fresh bad news: the commander of the guards rocket launcher units had received orders to take over the captured German rocket technology. In August, Gen. N.N. Kuznetsov arrived at Nordhausen and declared that the institute and all its employees were subject to the orders of the main artillery administration. The aviation industry had thus lost the Rabe Institute.

Soon there were talks with Gen. Lev Mikhailovich Gaidukov (head of the military council of guards rocket units and head of the corresponding department of the ZK). He called for the continued development of the institute.

After his return to Moscow, Gaidukov sent capable personnel to assist the institute: Mikhail Sergeyevich Ryasanski, Viktor Ivanovich Kuznetsov, Yuri Alexandrovich Pobedonostsev, Yevgeni Yakovlevich Boguslavski, and Sinoviy Moiseyevich Zezior (all future chief designers). Capable people were also sent from the artillery main administration.

German rocket experts, however, were still scarce. Since the October Revolution the German

Lev Mikhailovich Gaidukov.

people had been inoculated with hatred for bolshevism, and Goebbels' propaganda had made this worse. The mistrust ran deep. Some of the people had been uprooted, torn from their places of work, their companies destroyed, far from their homes. A new recruiting campaign was begun and it was quite successful. Those rocket experts prepared to cooperate were offered conditions, which, after the deprivations of the war years, were fantastic: high salaries, apartments, special food. Among the experts who were recruited were Dr. Hoch (automatic controls), Dr. Blasig (actuators), and Professor Wolf (formerly head of ballistics at Krupp).

Feelers were also extended into the American and British zones in an effort to recruit rocket experts. One particularly impressive success was the recruitment of the entire Gröttrup family. Helmut Gröttrup was von Braun's deputy for rocket radio guidance and electrical systems.

Dr. Kurt Magnus (theoretician and engineer for gyro theory) was convinced by a Dr. John, and on April 25, 1946, he signed a contract with the Gröttrup office.

The situation for families of the rocket experts in Moscow was far from rosy: there was a shortage of apartments, little food, now and then no water, the cold, childhood illnesses. It was thus

Quarters of Lt. Gen. Lev Mikhailovich Gaidukov, director of the central factory, as they appeared in 2010.

Helmut Gröttrup in 1945 (left) and Dr. Kurt Magnus (right).

understandable that the institute staff were eager to get to Moscow. Gaidukov did the only correct thing in that situation. He imposed a ban on leave for rocket experts working in Germany. The families in Moscow received support. The number of rocket experts assigned to the institute rose further, but the big problem was—there were no responsibilities for the rocket technology and the personnel. Gaidukov therefore decided to speak to Stalin. As part of his preparations he studied the exchange of telegrams with Churchill concerning the V-2. He studied the history of the RNII—the Rocketry Science Research Institute and its employees Korolev, Glushko, and others. With a list of names he facilitated the release of many rocket experts from Beria's special prisons. The conversation with Stalin went satisfactorily but without a ruling. Gaidukov was personally authorized to speak with the people's commissars about assuming responsibility for the development of rocket technology. Shakurin, People's Commissar for the Aviation Industry, refused. He was under pressure because the USSR was lagging behind in the development of jet aircraft. Vanikov, People's

Commissar for Munitions, also refused. He was also leader of the first main directorate involved with the nuclear weapons problems. This left Ustinov, the People's Commissar for Armaments. He thought about the perspective of his future ministry and concluded that, as a provider of weapons, he would always play second fiddle after tanks, aircraft, and ships. After the provisional concession allowing his people's commissariat to assume responsibility for "controlled rockets," he gave his deputy Vasiliy Mikhailovich Ryabikov the task of studying the problem in depth.

Although the use of V-2 rockets against London had failed to achieve the anticipated results, the military men of the guards rocket forces were convinced that the weapon had prospects. The calls by various authorities to cease the activities at Nordhausen and Bleicherode and move them home were blocked, and following tense negotiations the work at the Rabe Institute continued.

Ryabikov therefore soon went to Bleicherode. He visited the Mittelwerke, busied himself with the organization of the Rabe Institute, and listened to presentations on the development of the A-4 missile and Peenemünde. Finally, possible rocket development projects were discussed. Ryabikov was full of praise and promised to make detailed report to Ustinov.

Lev Mikhailovich Gaidukov played an important part in the development of Soviet rocket technology. It is impossible to say if Soviet rockets would exist today without his contributions.

It was an unwritten rule that the names of people repressed by Stalin were taboo. One could only talk about them at closed party gatherings, and in doing so one had to leave out the fact that one had once worked alongside an enemy of the people. The good manners of the time required that everyone who spoke denounced the enemy of the people, and in the eruption of self-criticism that followed listed every possible error in the work of the group, department, or the entire institute. Then one swore the great truth of Stalin, who just in time had ordered the intensification of the uncompromising class struggle. Then one had to declare that one would rally round the "big thing," correct the mistakes that had been made, stiffen the ranks, and achieve and even surpass the objectives. Then one had to forget the name of the enemy of the people. Books and articles by authors had to be removed from libraries and were placed in special storage. This taboo remained in place virtually until Stalin's death.

This was the situation when Chertok and Korolev met each other for the first time at the Rabe Institute. Gaidukov had kept his word and was able to free many rocket experts from camps and prisons.

After five years in prison and camps, Korolev was out of the picture, and now he had the freedom to work and make plans: which launch technology do we need? Who is responsible for launch preparation? Who is to organize the launch?

Days later Chertok and Pilyugin finally learned who Korolev was and that Gaidukov had given him the task of establishing an independent department. Its goal was to study the launch process in the following areas: launch preparation of the rockets, propellant and launch equipment, guidance technology, computation of the flight plan and flight parameters, instruction of the ground crew and the required documentation. Korolev's Launch Group was joined by a number of officers from the Rabe Institute, including Voskresenski, Rudnitski, and others.

The most important question that remained to be answered was: who has overall responsibility for all rocket technology in the country? At that time everything was still in the hands of the military.

How her father led the British around by the nose seemed to give Korolev's daughter a great deal of pleasure. In mid-October 1945, the British invited their brothers in arms to attend a demonstration launch of a V-2 at Cuxhaven. The operation was dubbed "Backfire." Korolev was ordered to Berlin, where he was made a member of the official division delegation under Gen. Sokolov.

All of the military men wore their regular uniforms, and orders from the highest level directed Korolev to appear in the uniform of an artillery captain. There was no end to the secrecy. British intelligence eyed him very attentively. An Englishman who spoke excellent Russian asked him about his lack of decorations: "You obviously weren't at the front?" The masquerade had leaked out, but the launch went ahead anyway. The British demonstrated a total lack of under-

Prof. Natalya Koroleva, daughter of Sergei Korolev, during her speech to the 24th Day of Spaceflight in Neubrandenburg.

standing as to what was taking place, and the German launch crew handled everything. Fog made it impossible to follow the flight of the rocket, but the launch was impressive.

What was the engine production situation? After the bombing of Peenemünde in August 1943, the Germans had sought locations for the underground production of A-4 rockets.

In the town of Lehesten near Saalfeld, known for its slate mining, the slate pits were examined and found suitable. The SS took over four miles of tunnels and fifty-four mining areas.

Beginning in 1943, test stands for the A-4's rocket engines were built in Oertelsbruch. The engines were assembled in the Mittelwerk Dora and transported by rail to Lehesten for testing.

The engines were tested on a firing stand. The number of stands was increased to two and later three. An underground oxygen plant was built that included a coolant system for liquid nitrogen, a propellant dump, a set of tracks, and other installations. Approximately thirty engines were tested per day. Production errors were corrected during testing, then the engines were preserved and sent back to Mittelwerk Dora and installed in the completed rockets. The engine test facility was integrated into the "Central Plant" under the title ZW 8 Lehesten.

The Oertelsbruch slate quarry was in "Happy Valley" near the present suburb of Schmiedebach. In September 1943, "Laura," a satellite camp of the Buchenwald concentration camp, was established with 209 inmates. Their number rose to 1,227 by the end of the war.

On July 15, 1945, engine rocket experts Alexey Isayev and Arvid Pallo travelled to Lehesten, where they easily found the former head of the engine-testing base. The Soviets could not understand why the Americans had not taken the German rocket experts and technology with them when they left. In addition to the engines, the turbopump assemblies were tested there and then delivered separately to the Mittelwerk. The engines and turbopump assemblies were then brought together during assembly—a technology that had proved less than satisfactory.

Arvid Pallo became "head of testing" and began organizing the engine testing program. An interesting discovery was made in an underground dump: more than fifty combustion chambers ready for testing. Fifteen rail cars arrived from Peenemünde carrying engines for the A-4, Meillerwagen trailers for transporting the rockets, tank cars for transporting liquid oxygen, and other useful equipment.

More then forty tests were carried out by the end of September. The Soviets had to learn how to break in and test the rocket engines. How could the operating conditions be changed? To the amazement of the Germans, the Soviet rocket experts succeeded in raising the thrust limit to 77,160 pounds. The V-2 produced 55,100 pounds of thrust. Work on the test stands became a unique practical-scientific effort. Technologies were developed to measure thrust parameters, to calculate and select the aperture, to measure the oxygen nozzle flow stream, an express method of analysis to determine the chemical and physical qualities of the propellant for the combustion chambers and the components of the steam generator, and to measure the alcohol nozzles' flow stream.

A few days later Glushko and List arrived in Bleicherode. After taking some time to rest, Glushko took charge in Lehesten. The work was given new stimulus. The engine-testing technology was developed further. Statistical data was gained about the burn processes, temperature development, and consumption characteristics. The working group also benefited from the discovery of additional rail cars with combustion chambers, launch equipment, transport equipment, and tank cars.

In the beginning the test firings were directed by Dr. Joachim Umpfenbach, a former propulsion expert from Peenemünde. In September, V.L. Shabransky became director of the test program and continued in this position until Soviet work on

This image is a still taken from the previously unknown film showing the test of a V-2 rocket completely carried out by Russian specialists in the Oertelsbruch near Schmiedebach.

rocket technology in Germany ended in January 1947. A total of 407 test ignitions were carried out by the combined team of German rocket experts and engineers from Glushko's OKB-456.

While research into German rocket technology continued at the Rabe Institute, in Moscow the institute's situation was less than rosy. Significant elements of the party and state apparatus responsible for Germany called for the immediate cessation of work there and the return of the rocket experts to the USSR by January–February 1946, at the latest.

At the beginning of January, however, Gaidukov, Ustinov, and Marshall Yakovlev of the artillery, who backed them, succeeded in convincing those in power to continue and expand the work, on the condition that an even more capable organization be created.

The Rabe Institute was obviously set up to investigate the problems relating to the electronic guidance system. Chertok, Pilyugin, Rosenplenter, Dr. Hermann, and Gröttrup were rocket experts in the field of electronics and guidance systems. Korolev headed the "launch" group. Glushko studied the rocket engines at Lehesten. Kurilo oversaw the assembly of rockets in Kleinbodungen. There were also a small number of other groups that worked more-or-less independently. This certainly could not be described as an organizational unit.

Korolev was summoned to Moscow in February 1946. At the end of March, he returned to Bleicherode as a colonel and thus held the same rank as Glushko, Pobedonozev, Ryasanskiy, Pilyugin, and Kuznetsov. Two days later a beaming Gen. Gaidukov also arrived

from Moscow. At a gathering of Soviet rocket experts he announced the creation of the Nordhausen Institute, with the objective of combining all of the groups that were working separately into a single organization. It was headed by Gaidukov, and his first deputy and chief designer was Korolev. He also announced the arrival of additional rocket experts.

The new structure looked like this: the directorate of the Nordhausen Institute was established in Bleicherode.

• The Rabe Institute was incorporated into the structure as the institute for guidance systems. The old leadership remained, but the number of staff was reduced.

• The installations in Lehesten and the Montania factory near Nordhausen for the production of engines and turbopumps were combined under Glushko. Shrabanski was named to command at Lehesten.

• The operation in Kleinbodungen was named Factory 3 under Kurilo. Its job was to reconstruct the production technology and manufacture the maximum possible number of rockets from the remnants that had been found.

• The Olympia design bureau in Sömmerda. Its task was reconstruction of the A-4's documentation and the technological equipment. It was led by Budnik and then Mishkin. Documents discovered in Prague simplified the start of work.

The savings bank in Bleicherode in 2010—this is where the computer and theoretician group was accommodated.

• The computing and theoretician group, called the "Savings Bank," was housed in the bank. Tyulin had control, assisted by Lavrov, Moeshorin, Apasov, theorists from the Rabe Institute, Dr. Wolf (head of ballistics at Krupp), and the aerodynamicist Dr. Albring.

• The Gröttrup Office was incorporated as an independent department. One of its most vital tasks was expanding the history of the development of the A-4 rockets at Peenemünde.

• The Launch Group. Vokresenskiy replaced Korolev as head of the group, whose staff was significantly enlarged. Its task was to search for and reconstruct propellant and transport technology.

Also established at this time was the Berlin Institute. Barmin, its chief engineer, was the future designer of ground equipment for rocket launches.

Despite his new position, Korolev continued working with the Launch Group. With Mishkin and Budnik he began the first preparations for development of a rocket with a range of miles, the later R-2.

In general the German rocket experts were well provided for; however, two problems could not be eliminated. There was a certain mutual mistrust. The Soviet rocket experts suspected that the Germans were secretly carrying out sabotage, while the Germans feared for the future. What will become of us when the Russians have learned everything and don't need us any more? The work was continued and expanded, and staff figures rose from 6,000 to 7,000 people.

B.E. Chertok and S.P. Korolev in Germany.

ORDER FROM THE COUNCIL OF MINISTERS

After the first launch of a V-2 in the desert of New Mexico, the Soviet government saw itself forced to act. On April 29, 1946, from 2100 to 2245, a consultation took place in Josef Stalin's Kremlin cabinet on the questions of rocket construction and the use of rockets as weapons. The consultation was the result of a written report by Stalin dated April 17, and bearing the signatures of L.P. Beria, G.M. Malenkov, N.A. Bulganin, B.L. Vanikov, D.F. Ustinov, and N.D. Yakovlev.

At the consultation the following topics were addressed:

• About further work by the institute located in Nordhausen and the design and technological bureaus for the V-2 organized in Germany, and also about the experts working there; about the extent of the transfer of equipment and German specialists working on the V-2 in Germany to the USSR.

• About the organization of the experimental V-2 test rockets gathered at Nordhausen; about the necessity of creating a special testing site.

The results were reflected in Council of Ministers resolution 1017-419 of May 13, 1946. The resolution by the Council of Ministers was secret.

Council of Ministers of the USSR
Directive
No. 1017-419ss of May 13, 1946
Matters of Reactive Equipment

Including the most important task of creating reactive equipment and the organization of the research and experimental work in this area, the Council of Ministers of the SSR RESOLVES

I

1. To create the Council of Ministers of the SSR's Special Committee for Reactive Technology:

Gen.* Malenkov, G.M. [1] – chairman
Gen. Ustinov, D.F. – deputy chairman
Gen. Zubovich, I.G. – deputy chairman, relieved of his work in the Ministry of the Electric Industry
Gen. Yakovlev, N.D. – member of the committee
Gen. Kirpichnikov, P.I. – member of the committee

Gen. Berg, A.I. – member of the committee
Gen. Goremykin, P.N. – member of the committee
Gen. Serov, I.A. – member of the committee
Gen. Nozovski, N.E. – member of the committee
* stands for "Comrade," usual form of address.

2. To assign the following responsibilities to the Special Committee for Reactive Technology:

a) Monitor the development of research, design, and practical work related to reactive armament, immediate consideration, and submission to confirm plans and the development program for research and practical work in the named fields, and also the determination and confirmation of quarterly requirements for financial allocations and material-technical resources for work on reactive equipment by the chairman of the Council of Ministers of the USSR;

4. Ascertain that the work on reactive armament carried out by the ministries and offices is overseen by the Special Committee for Reactive Technology. No institutions, organizations, or persons have the right, without a special decision by the Council of Ministers, to interfere with or make demands of the work on reactive armament.

II.

6. Designated leading ministries in the development and production of rocket technology are:

a) The Ministry of Armaments for liquid-fueled rockets,

b) the Ministry for Production of Agricultural Machinery for solid-fuel rockets,

c) the Ministry of the Aviation Industry for winged rockets.

7. The committee designates as basic ministries for subsequent production, to take over orders for execution of scientific research, design, and experimental work, likewise production on behalf of the leading ministries confirmed by the committee:

a) the Ministry of the Electric Industry for ground and onboard radio guidance systems, selected apparatuses and television mechanisms, radar systems for tracking and the determination of target coordinates,

b) the Ministry for Shipbuilding for gyro-stabilization systems, computing equipment, sea-based tracking stations, self-guided warheads for use against underwater targets, and equipment,

c) the Ministry for the Chemical Industry for liquid fuels, oxidizers, and catalysts,

d) the Ministry of the Aviation Industry for liquid-fuel rocket propulsion for long-range combat rockets and production for aerodynamic research and rocket testing,

e) the Ministry for Mechanical and Apparatus Engineering for installations, launch facilities, various compressors, pumps, and apparatuses for this and other completion apparatuses,

f) the Ministry of Agricultural Mechanical Engineering for contactless igniters, explosives, and powders.

III

8. A reactive armament department is being created in the state plan under the direction of the deputy chairman of state planning.

9. The following scientific research institutes, design bureaus, and rocket launch sites are to be created in the ministries:

a) in the Ministry of Armaments - a scientific research institute for rocket weapons and a design bureau on the basis of Factory No.88, which is to give up all other tasks, which will be carried out by other facilities of the Ministry for Armaments;

b) in the Ministry of Agricultural Mechanical Engineering - a scientific research institute for solid-fuel rockets on the basis of GZKB-1, a design bureau of Branch No.2 of the Ministry for the Aviation Industry's NII-1 and a scientific research test range based on the firing range at Sofrony;

c) in the Ministry of the Chemical Industry - a scientific research institute with project design bureau for chemicals and rocket fuels;

d) in the Ministry of the Electronics Industry - a scientific research institute for radio and electronic devices for guidance of long-range offensive and anti-aircraft missiles based on the laboratory for telemechanics NII-20 and of Factory No.1.

e) in the Ministry of Armaments - the scientific research institute of the Main Administration Artillery and the State Central Test Range for Rocket Technology for all ministries involved in rocket technology.

IV

11. We consider it a primary task to carry out the following work on rocket technology in Germany:

a) the complete reconstruction of the technical documentation and design of the V-2 long-range guided missile and the guided air defense missiles – Wasserfall, Rheintochter, Schmetterling, and other missiles.

b) the reconstruction of laboratories and test stands with all equipment required for the necessary investigation and testing of V-2 missiles, Wasserfall, Rheintochter, and Schmetterling and other missiles.

c) the preparation of cadres of Soviet experts, who are to master the design of the V-2 rocket, the air defense missiles, and other missiles, the [German] experimental methods, the technology for production of parts and systems, and the assembly of the missiles.

12. Comrade Nozovsky is named head of rocket technology in Germany with his residence in Germany. Comrade Nozovsky is to be relieved of other tasks not associated with rocket armaments. Generals Kuznetsov (Main Administration Artillery) and Gaidukov are named Comrade Nozovsky's deputies.

13. The Committee for Rocket Technology is obligated to select the necessary number of specialists from different fields from the various ministries and to send them to Germany to study and to work in the field of rocket technology. To achieve the goal of gathering sufficient experience, each German expert must be assigned Soviet specialists.

14. The ministries and authorities are forbidden from recalling their employees who are in Germany and working in committees to research German rocket weapons without the approval of the special committee of their coworkers.

16. The Ministry of Defense (Comrade Bulganin) is mandated to form a special artillery battalion in Germany for the mastery, the preparation, and the launch of V-2 rockets.

17. The question of transferring the design bureaus and the German experts from Germany to the USSR by the end of 1946, has already been decided. The Ministries for Armament, Agricultural Mechanical Engineering, Electronics Industry, Aviation Industry, Chemical Industry, and Mechanical and Apparatus Engineering are obligated to prepare bases for housing the German design bureaus and specialists.

19. The Ministry of the Armed Forces of the USSR (Gen. Krulev) is obligated to provide for the supply of all Soviet and German specialists working on reactive armaments in Germany:

- free shares as per standard No.11
- 1,000
- as per standard No.2 with additional shares
- 3,000
- automobiles – 100
- trucks – 100

Also to supply fuel and driver's station.

20. The Finance Ministry and the Soviet military administration in Germany are to be obligated to provide seventy million Marks for the financing of all work carried out by the special committee for reactive armaments in Germany.

21. To allow the special committee for reactive armaments and the ministries to order special equipment and apparatuses for the laboratories of the research institute and of the state central rocket range in Germany as part of reparations. The special committee together with the state plan and the Ministry of Foreign Trade are to be consulted, to determine the list of orders and the dates of their delivery.

V

25. To direct the Ministry of Armed Forces of the USSR (Comrade Bulganin) to submit a proposal to the Council of Ministers concerning the location and construction of the state central rocket range.

26. To obligate the Minister of Higher Education Comrade Kaftanov to organize the preparation of engineers and scientists for reactive technology in the highest learning institutions and universities, but also to retrain students from earlier courses in other fields in reactive equipment. It is to be ensured that the highest technical learning institutions provide no fewer than 200 people and the universities no fewer than 100 people as the first specialized graduates for reactive equipment by the end of 1946.

31. With the goal of accommodating the German specialists in rocket technology in the USSR, Comrade Vosnesenski is directed, as per the envisaged distribution plans, to provide the Special Committee for Rocket Technology with 150 collapsible small houses and forty wooden eight-room houses by October 15, 1946.

32. Development of reactive technology is to be regarded as the most important government task, and all ministries and organizations are obligated to complete their tasks in support of reactive technology as a matter or urgency.

The chairman of the Council of Ministers of the USSR,
J. Stalin

The managing director of the Council of Ministers of the USSR,
J. Chadayev

Responsibilities in Soviet Rocket Construction (Council of Ministers Directive No. 1017-419ss of May 13, 1946)					
Ministry	Minister	Institute/ Design Bureau	Location	Area	Chief Designer/ Director
Armament	D.F. Ustinov	NII-88	Kaliningrad	Main client	L.R. Gonor
		OKB-1	Gorodomlya		S.P. Korolev
		OKB-2	Zagorsk		K.I. Tritko
Aviation Industry	M.V. Krunichev	OKB-456	Khimki	Engines	V.P. Glushko
Mechanical and Apparatus Engineering	P.I. Parshin	GSKB	Moscow	Launch complex	J.V. Barmin
Signals Technology	I.G. Subovich	NII-885	Monino	Guidance systems	M.S, Tyazanski
Electronics Industry	I.G. Kabanov	NII-20	Moscow	Telemetry	Chistyakov
Agricultural Mechanical Engineering	P.N. Goremykin	NII-1	Vladykino	Heating processes	M.V. Keldysh
Shipbuilding	A.A. Goreglyad	NII-10	Moscow	Gyroscopic devices	V.I. Kuznetsov
Chemical Industry	M.G. Pervukhin	GIPCh	Leningrad	Fuels	P.L. Prokofyev

The Council of Ministers of the Union of Soviet Socialist Republic's special committee for reactive technology faced a mountain of tasks:

- Acquisition of German rocket technology
- Formation of a rocket brigade
- Construction of a rocket launch site
- Qualification of several thousand rocket experts
- Relocation of German rocket experts to the Soviet Union by the end of 1946
- Expansion or conversion of businesses

For the Soviet reengineering of the V-2 alone, the production of sixty-three factories under the heads of sixteen different ministries had to be coordinated. Almost all of the businesses and institutes created at that time still exist today, even if under different names.

One can say that the A-4 rocket resulted in the development of the first complex system in the Soviet Union—a tremendous material, personnel, and financial challenge. As a result of the work carried out in Germany, the Soviets were forced to realize what an extraordinarily complex and multi-faceted technical task the development of a ballistic missile system was. Many special installations tailored to this specific rocket were required: launch equipment, guidance systems, and other safeguard systems. These requirements could only be met by exploiting the scientific and industrial potential of the entire country and the efforts of many specialized research, design, and production collectives, which had thousands of highly qualified rocket experts in all areas.

Simultaneous with the order from the Council of Ministers of the USSR on May 13, 1946, it was decided to form a Council of Chief Designers. On the initiative of S.P. Korolev, six chief designers formed the Council of Chief Designers, which had

to make decisions on all fundamental questions concerning rocket technology. The original members of the council were:

- Sergei Pavlovich Korolev: Chief designer of OKB-1 (Main Design Bureau 1) of the total rocket system.
- Valentin Petrovich Glushko: chief designer of OKB-456 for liquid-propellant rocket engines.
- Vladimir Pavlovich Barmin: chief designer of the GSKB (specialized guidance design bureau) Spezmasch for launch, transport, and tank equipment.
- Nicolai Alekseyevich Pilyugin: chief designer of NII-885 (scientific research institute) for independent guidance systems.
- Mikhail Sergeyevich Ryazanski: work on autonomous guidance systems; from February 1947, chief designer of Scientific Research Institute 885 for radio communication equipment for the rockets and radio navigation.
- Viktor Ivanovich Kuznetsov: chief designer of the Scientific Research Institute of the Shipbuilding Industry for gyroscopic control units.

Council of Chief Designers: A.F. Bogomolov, M.S. Ryanskiy, N.A. Pilyugin, S.P. Korolev, V.P. Glushko, V.P. Barmin, V.I. Kuznetsov (1954).

This group is sometimes called "the big six." In Germany the future creators of the rocket-space branch of the Soviet Union got to know each other and there began working together for the first time.

Each of these chief designers was also responsible for a design bureau and the businesses linked to it. This kind of organization ensured that the decisions made within this small circle were effective on a large scale. This kind of decision-making was successfully used both by the military (war councils) and in industry (shipbuilding).

The order of May 13, 1946, also laid the foundation for the first Soviet rocket brigade, which was formed the following month. The Special Purpose Brigade of the High Command Reserve was formed on the basis of the 92nd Guards Rocket Launcher Brigade. Gen. Aleksandr Fedorovich Tveretskiy was named as its first commanding officer. The unit was stationed at Berka, 1.8 miles east of Sondershausen in Thuringia.

The village of Berka (symbols A to C) near Sondershausen.

Korolev, Voskresenski, Pilyugin, and Chertok drove from Bleicherode to Sondershausen, where the brigade's officer corps was quartered. Fears that rocket technology could fall into the hands of inexperienced commanders and bring discredit on the whole project proved unfounded.

Tveretskiy was both intelligent and assertive. The officers and military rocket experts now arriving daily were sent to the various laboratories and departments of the Nordhausen Institute to become familiar with the diverse aspects of rocket technology. It should be remembered that many of the officers were experienced combat officers who had to work their way into an entirely new field. There was little fluctuation. On the contrary: not only did the officers remain true to the rocket, they also decisively influenced the development of rocket technology. Several examples:

- Nikolai Mikolaevich Smirnitsky: in 1955, head of the Main Administration for Rocket Weapons, later deputy senior commander of strategic rockets.
- Yakov Isayevich Tregub: chief designer Mishkin's deputy, worked on robotic cosmic devices.
- Aleksandr Ivanovich Nozov: led testing of the R-7 intercontinental missile at Baikonur.
- Boris Aleksandrovich Komisarov: deputy head of the military-industrial committee of the presidium of the Council of ministers of the USSR.

As usual in many such cases, no one knows where the idea came from and all were for it. So at the beginning of 1946, the idea was born to develop and build a railroad-based special rocket train. As the Nordhausen Institute's resources were limited, the contract for the construction of the cars and delivery dates was taken care of by SMAD. The purpose of the train was to test and launch rockets from selected sites. The only requirement was tracks. The train was to consist of more than twenty special cars and flatcars. The concept included:

- Laboratory cars for autonomous checking of all equipment on the rocket
- Messina radio-telemetry cars
- photo laboratory for developing film
- cars for testing the engines' automatic system and fittings
- generator cars for producing electricity
- compressor cars
- workshop cars equipped with workbenches
- a medical car
- dining cars
- five comfortable living cars with two-seat compartments

- cars with baths and showers
- salon cars for holding meetings
- two salon cars for the senior leadership
- armored cars with electrical firing systems

The rockets were to be fired from the armored rail cars. The rocket was placed on the launch platform, which together with a loading device was part of the special flatcar. Under such conditions it was possible to work without turf huts and tents, making the crew more or less able to work in any weather.

The building of this train must have left such a deep impression on brigade commander Tveretskiy that he obtained authorization and the necessary funding for the construction of a second special train just for the military. Doubling of the program led to numerous conflicts, because there was not enough special testing and measuring equipment with which to equip the cars.

Despite everything both trains were completed and completely fitted out by December 1946.

Work at the Nordhausen Institute expanded. The rocket parts that had been found, checked, and reconstructed were gathered in Kleinbodungen "Factory Three" and assembled into working rockets. The engines had previously been checked out at Lehesten and the turbopumps came from the Montan factory. In August, rather more than twelve relatively complete V-2s were stored there.

There were, however, considerable shortfalls in the measurement and testing technology. The Carl-Zeiss Company in Jena proved to be the knight in shining armor. All of the products ordered were delivered by September. Other parts were organized by German employees during business trips to the western zones.

And once again the Nordhausen Institute was visited by a high-ranking government delegation. It was led by Marshall of the Artillery Yakovlev. Other members of the committee were:

- Col. Gen. Dmitrii Fedorovich Ustinov, Minister of Armaments

- Maj. Gen. Lev Robertovich Gonor, named director of NII-88
- Col. Sergey Vetoshkin, head of the main administration of the Ministry of Armaments
- Georgi Nikolaevich Pashkov, head of the department for the defense industry of the State Planning Commission
- Voronzov, deputy of the Minister of Radio Industry

Ustinov advised that his ministry was being made a senior ministry and that Korolev was working with him as chief designer. As a result of the decision by the Council of Ministers of the USSR on May 13, 1946, many responsibilities in the area of rocket technology had already been assigned, but many questions concerning project planning, production, and especially the question of cares remained open for the future.

For the present, there were still discussions as to what part German rocket development would play, in Soviet rocket development, for example.

At a meeting Minister Ustinov expressed the following opinion:

"Great and important work was done here (in the Nordhausen Institute). Our industry did not have to start from zero, from an empty patch. But we will first learn what was created here in Germany. We will exactly reproduce the German technology before we begin creating our own. I know that this does not please some people. They too have found major shortcomings in the German rockets and are inspired by the desire to follow our own path. But for the first time we are strictly forbidden from doing so. First they must prove that they can do it just as well. And those who summon our experience and history, to them I answer quite clearly, we have the right to it, for we have paid for it in blood!

But we are forcing no one. Anyone who does not wish to play a part can look for other work."

For development and production of the engines, it as decided that German engine production would not be used. A special factory was being built for Glushko in Khimki.

The guidance systems were being developed by the radio industry under the leadership of Ryazanski. His deputy was Pilyugin.

Development of the actuators stayed at the main institute. Chertok was to go to NII-88 as Pobedonostsev's deputy, meaning as chief engineer.

The personnel questions were thus taken care of. Aviation had lost its grip on rocket technology and Ustinov was in charge.

Division of the laboratory facilities and production inventory was carried out without problems. It was agreed that the creation of documentation would take place in such a way that each would receive what was necessary. The originals, however—the copies and the copies of copies—had to be held in NII-88's central archive.

Gaidukov and Korolev received orders to produce a detailed report about the activities of the Nordhausen Institute. In conversation it was also mentioned for the first time that work in Germany would be ended at the end of 1946.

Gonor demanded that Pobedonostsev, Korolev's deputy for technical documentation, and a number of other leading rocket experts arrive in Podlipki no later than September. This resulted in Soviet rocket experts beginning to leave Germany in August, and settling in the area around Moscow.

But there were still problems that had not been solved. The Moscow laboratories lacked equipment that had been installed in the special rocket trains. Equipment had to be found for two special laboratories, in particular various test stands.

At this time the assembly of twelve rockets and their horizontal testing was also completed. The horizontal testing proved to be a complicated technological process that was to give the Soviet rocket builders problems for years. A high level of technical knowledge was required in order to quickly find faults. Of the rockets checked out, none had fewer than ten faults. Contacts, failed equipment, and indicator lights were frequent sources of trouble. Troubleshooting! This demonstrated clearly the poor reliability of the A-4's electrical switching system.

Individual assemblies were also put together. Then they were supposed to be assembled into a complete rocket in Podlipki, a teaching tool for the further qualification of the employees.

S.P. Korolev in one of the last photos taken in Bleicherode, 1947.

DEPARTURE AND THE END

By order of Lt. Gen. Serov, Beria's deputy, at the beginning of October 1946, all of the important heads at the Nordhausen Institute met in Gaidukov's office. There they were told to compile lists and brief biographies of German rocket experts who were to work in the Soviet Union. The consent of the rocket experts was not required.

Serov declared: "We will allow the Germans to take everything, including furniture, because we ourselves have problems with this. As far as family members are concerned, this depends on the desires of the individual families. If the wife and children wish to go along, they are welcome, if they don't, then they are not. They don't need to do anything except take part in a farewell banquet. Give them enough to drink, that will make the whole business easier to accept. No one will be told about this decision, to prevent the rocket experts from disappearing. A similar operation will be carried out in Berlin and Dessau at the same time."

It is difficult to determine how the Soviet rocket experts felt about working with their German colleagues. After endless discussions during the afternoon and evening, the participants were taken by bus to the "Blecherode Kremlin," the Japan Restaurant. The table was filled with good food, and there was also plenty of vodka.

There was toast after toast. By chance, however, some Germans discovered that the Soviet officers were only drinking water. Leaving the room was not possible. The intention was to get the German transport candidates drunk.

Four in the morning, October 20, 1946, hundreds of military trucks drove through the town. Each designated representative sought out a house. The interpreter rang the bell and, after a brief conversation, the order, which read roughly as follows, was read out:

"We must inform you that the Soviet Ministry of Armaments has ordered the central factory moved to the Soviet Union. The factory's rocket experts, including you, are therefore to work in the Soviet Union in the coming years . . . Some members of the German work force, in keeping with the reparations obligation in the Potsdam Agreement, are required to work in the Soviet Union. . . . Departure commences today."

Arguments, resignations—it made no difference. All counterarguments were listened to cordially but rejected. Two thoughts entered the heads of many Germans—kidnapping and Siberia.

Beginning at noon, under the supervision of the respective masters of the house, the personal belongings were loaded into trucks, driven to the

Entrance to the Japan Restaurant and a view of the dining room.

Bleicherode in the year 2010.

Kleinbodungen or Bleicherode railway stations, and placed into boxcars. The people boarded passenger cars. A total of sixty rail cars had been prepared. Large areas around the stations were sealed off and illuminated.

The only serious problem involved the Gröttrup Family. *Frau* Gröttrup insisted that her two cows must go with her. They could not allow their two children to starve. This problem, too, was solved by adding a car with hay.

There was a constant coming and going. Items that were forgotten were returned for. Workers who were still active outside town were brought in. In the evening everyone was asked to gather in a shed. The officer in charge of the transport read out the order again. Many questions were asked, few were answered. One thing had become clear, however; the train was going to Moscow. They were obligated to work for two years, after which all of the rocket experts would be returned home.

Under these conditions sleep was out of the question. Each was left to his own thoughts and was unable to rest. Finally, on the afternoon on October 23, the train began to move, toward an uncertain fate.

The attitudes of the German rocket experts could not have been more different: depression, indifference, anger, protest. Many were in a traumatic state. There was also breach of contract. Some would have voluntarily worked in the Soviet Union if they had been asked. There were also volunteers whose names were not on the long lists.

The operation was carried out with the necessary secrecy and took those affected completely by surprise. Even those involved later spoke of an exemplary organization. In Bleicherode and surroundings between 150 and 180 persons and their families were affected. The known personalities and their areas of involvement:

Helmut Gröttrup	head of the group
Dr. Waldemar Wolff	ballistics
Josef Blass	design
Dr. Franz Mattheis	chemistry
Dr. Erich Apel	workshops
Dr. Werner Albring	aerodynamics/projects
Dr. Joachim Umpfenbach	engines
Dipl.-Ing. Rudolf Müller	statics
Dipl.-Ing. Heinz Jaffke	test stands
Prof. Wilhelm Schützmetrology	
Dr. Hans Hoch	guidance systems
Ing. Fritz Viehbach	launch organization
Prof. Walter Pauer	thermodynamics

Also, doctors John, Zopf, Döhring, Sulzer, Stöger, Thamm, Baldung, Magnus, Professor Doctors Thiels, Vulpius, Malter, Gerber, Reschke, Nolte, and rocket experts Kalker, Balke, Chestmann, Hans Kade, Nürnberg, Rauhe, Nehrkorn, Pehle, Rühdiger, Scholz, Stahl, Vilter, Wollfahrt, and Töpfer.

Not only were the rocket technicians from Bleicherode affected, but also a total of 5,000 rocket experts from the various specialist areas in eastern Germany—Jena, Berlin, Dresden, and Dessau. At least ninety trains were put together, and within five days they all left Germany headed for the Soviet Union. The rocket experts were to help rebuild the wrecked Soviet economy according to their qualifications. For the German rocket experts it was something different, however. They had reached a dead point. The work done required a practical test, meaning a rocket launch; however, given the political conditions in Germany this was impossible. This realization was reflected in the order by the Council of Ministers of the USSR on May 13, 1946.

Only from this point of view can the different viewpoints of the affected German and Soviet rocket experts be understood: one group as kidnapped, deported, deceived; the other joy over the rapid acquisition of a new weapon in order to soon be able to have something with which to oppose the American atomic cudgel.

The departure of the German rocket experts meant that the end of the Rabe and Nordhausen Institutes was near. Thoughts of further transports of rocket experts to the Soviet Union were rejected. There was still plenty of work to do. At least three tasks had to be completed: reproduction and completion of the documentation, packing up of the laboratories and production equipment, and the receipt of materials and equipment ordered from suppliers. Outstanding accounts also had to be paid. Including the packing of personal effects, this all dragged on until early 1947. Deserving of special recognition is the special action group that organized and secured the transport of German rocket experts but also assisted with the procurement effort. The last members did not leave Thuringia for the central launch site, which was under construction, until the summer of 1947, taking with them several A-4 rockets.

Most of the Soviet rocket experts returned from peaceful, tranquil Thuringia in the spring and were confronted by the tough reality of postwar Moscow: four people, two rooms, no bath, no shower, one closet and a wash basin, a wood stove, drafty windows. Many workers were housed in hastily-erected barracks. Accommodations were sought with friends and relatives.

It was incomprehensible that after the victory over fascism, in which both those at the front and at home had given their utmost, that the system of repression struck back at the people with full force. Those who had been compelled to work as forced laborers, had been prisoners of war, or had simply lived in occupied areas had it particularly hard. The scientists dreamed of further cooperation with the scientists of the country's allies.

In 1947, the Baumann Technical University in Moscow was turned into a center for rocket technology. High-ranking rocket experts, like Korolev, gave lectures. Speaking about German techniques and experience was unavoidable, until one day word came from above that the word "Germany" was no longer to be mentioned.

What had been learned during the two years of work in Germany by Soviet rocket experts?

• The A-4 long-range rocket existed. It was no figment of the imagination.

• Rockets of this type could be mass produced. It had to be expected that in the future this rocket would be widely used.

• The rocket experts had the opportunity to get to know the design elements of the rocket, to test them, and to learn their strengths and weaknesses and thus draw conclusions for future developments.

• The broad examination of widest areas of German rocket technology had drawn the attention of the party and state apparatus of the USSR and of the military. This is confirmed by the order issued by the Council of Ministers of the USSR on May 13, 1946.

• The stocktaking and investigation of German rocket technology in Germany was correct. Some of the production sites could be inspected and put into operation. It proved possible to locate former scientists, employees, and workers and to convince them to work for the Soviets. The subcontractors also performed reliably and responded to the unusual requests made by the rocket experts, Working conditions in Bleicherode and other towns were much better than in the destroyed Soviet Union.

• During the existence of the Rabe and Nordhausen Institutes, many officers from the artillery and guards

Cutaway drawing of the A-4.

rocket units worked in their laboratories, test stands, and workshops. They gained extensive knowledge through their work and thus became the advance guard of the future rocket officers of the Soviet military and later held senior positions. Young scientists and technical school graduates also learned at the institutes. It was retraining and requalification. It is therefore not inappropriate to say that the Rabe and Nordhausen Institutes forged cadres for the future Soviet rocket industry and rocket forces.

• One needs qualified personnel to launch a rocket. This led to the formation of the first rocket brigade at Sondershausen/Berka.

• The Soviet rocket experts recognized that the rockets with their equipment and ground and launch equipment were much more complicated than previous weapons systems.

• The human side can also not be underestimated. The chief designers got to know one another. Scientists, military men, engineers—there were several thousand of them—worked together. Differences arose and were resolved. Friendships were made. Korolev put it thus: "The most valuable thing we achieved in Germany was building the foundation of a closed creative collective of like-minded people."

KAPUSTIN YAR

The scientific research institute NII-88 was located in the Moscow suburb of Podlipki. The V-2 rockets assembled in Thuringia and other equipment from Germany arrived there, likewise the special rocket train. The new rocket works was to be built there on the site of an old airfield. The first inspections could only be characterized with the terms "catastrophic" and "stone-age." Korolev and Ustinov made great efforts to adapt Soviet production culture to the requirements of rocket technology. The psychology of the workers and technologists also had to be improved. This took more time than envisaged in the plans and timetables.

Experience in Germany had shown: if a rocket is moved, meaning transported to another place, faults again arise. Great attention was therefore paid to simulation of the processes within the rockets.

NII-88 had a total of twenty rockets in two batches of ten. The N series had been assembled in Kleinbodungen, T series in Podlipki. The engines checked out at Lehesten, along with the turbopumps, were sent to Glushko at OKB-456 in Khimki, where they were checked again.

Gyroscopic devices and guidance equipment were sent to other installations, where they were checked, modernized, and also redesigned.

Inside the special rocket train.

In September 1947, everything was prepared for a launch. The testing group took the special rocket train to Kapustin Yar. In the first years of the rocket age, living and working at the test range in Kapustin Yar would have been unimaginable for the rocket experts. For it was not until the beginning of the 1950s, that hotels, buildings with workshops, and assembly and testing equipment plus recreation rooms were built on the initiative of Vasily Ivanovich Voznyuk, head of the State Central Test Range.

The decision to create a central rocket range was made pursuant to the order of the Council of Ministers of the USSR of May 13, 1946. A group led by Maj. Gen. Vasily Ivanovich Voznyuk was ordered to look for a suitable place to build the rocket range. Seven possible locations were chosen and in a very short time were carefully evaluated. Information about the following areas was gathered and checked: economic situation, weather, hydrology, communications, conditions for further development, building potential, and so on.

Based on the group's recommendation, the area of the village of Kapustin Yar in the Astrakhan area on the Lower Volga was chosen as the construction site for the rocket facility. The area was assessed unsuitable for economic exploitation—it was dusty, sunburned steppe. The only opposition to construction of the rocket launching facility came from Kozlov, the Minister for Livestock. He opposed construction of the rocket launch site because he feared that it would result in the loss of a significant portion of the grazing land.

It is interesting that in 2009, the location of the Kapustin Yar cosmodrome still differs from that of the actual village by 150 miles. Secrecy lives on.

The following was found in a written report by Marshall of Artillery Nikolai Yakovlev: "Construction

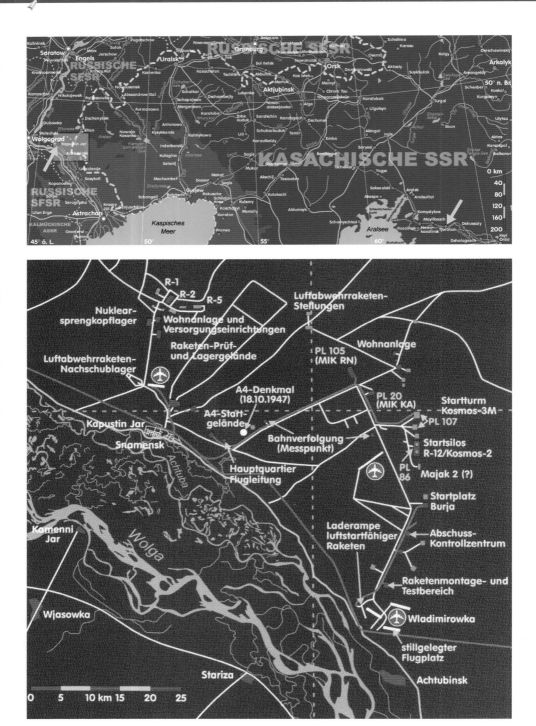

The position of Kapustin Yar; map basis and political-administrative designations from the years 1967 (top map) and 2009 (below).

Colonel General Vassili Ivanovich Vozynuk.

of the Central State Rocket Testing and Development site in the area of the Cossack settlement of Kapustin Yar makes it possible to create test ranges up to a distance of 1,800 miles and to ensure, not only the testing of long-range missiles, but also of all types of ground-based anti-aircraft and reactive projectiles for the navy. These variants will require the least material outlay for the resettlement of the local population to other areas."

Order ZKWKP (b) 2642-817 issued by the Council of Ministers of the USSR on June 3, 1947, determined the location of the central rocket testing and development facility of Kapustin Yar. The Kapustin Yar rocket range extended between 46 and 51 degrees east longitude and 65 to 67 degrees north latitude. V.I. Voznyuk, the energetic leader of the group that selected the location, was named the first commander of the rocket testing site.

July 23, 1947: at the decision of the Defense Minister of the USSR, a Department of Reactive Technology consisting of eight persons was created in the general staff.

On August 20, 1947, the first officers arrived at the testing site. Tents were pitched, kitchens and a hospital were organized. Along with Voznyuk's guards troops came military construction workers. Conditions were difficult, if one can say anything about any sort of conditions in the naked steppe. From the beginning, work was carried out with great enthusiasm.

On the third day, work began on the concrete stand for firing tests of the V-2's engines on the slope of the Smyslina Gorge six miles from the village. It was built from German plans and was equipped with the equipment brought from Germany. The bunker for observing the test runs also used equipment brought from Germany.

Col. Gen. Vassily Ivanovich Voznyuk (1907–1976) was born in the city of Gaizin in the Vinnitsa District, into the family of an actor. From 1923, he worked as a prompter and stage worker at various theaters.

In 1925, he volunteered to serve in the Red Worker's and Farmer's Army and was delegated to the 1st Leningrad Artillery School. After graduation in 1929, he was named platoon leader in an artillery regiment. In 1938, he became the regiment's chief of staff. From 1938 to 1939, he attended the Pensa Artillery School, after which he served in various staff and command positions in the active army in guards mortar units of the high command reserve in the Central Army, the Briansk, the Southwest, the 3rd Ukrainian, and the Voronezh Fronts. From June 1946, until November 1973, Voznyuk commanded the Kapustin Yar Cosmodrome. Under his command this facility was transformed into the largest testing and research center in the Soviet Union.

This place was later named Site 1. In September 1947, the rocket brigade under Maj. Gen. Alexander Fedorovich Tveretskiy arrived from Thuringia. Two special trains carrying equipment from Germany arrived somewhat later.

There was much building, only for the A-4 rocket, which was first on the list of priorities. The construction of living quarters for the personnel at the test site was postponed until 1948. The construction workers and future inspectors therefore lived in the barren steppe in tents, earth bunkers, and temporary buildings or quartered themselves in farmer's huts.

Construction of the concrete stand for burn trials with the V-2 engine.

Transporting an A-4 on a Meillerwagen.

On October 1, 1947, Gen. Voznyuk reported to Moscow that the testing site was completely ready to carry out rocket launches.

By the beginning of October 1947, after about one and a half months of work, apart from the concrete test stand and the bunker at the first site, the launch site with bunker, the temporary technical position, and the assembly hall had been constructed.

An avenue and a twelve-mile-long branch line with a bridge over a deep ravine were also laid down, linking the rocket testing site with the main artery to Stalingrad (Volgograd).

The leadership and the rocket experts who arrived at the rocket test site in September, lived in the special train Messina, which in addition to laboratories had very comfortable cars and even a dining car.

The first task was to place a rocket in a stand and carry out the complex operating test. The launch site was concreted and the launch platform was set up on it. While the military construction troops put up other installations, the rocket experts began equipping the facilities that had been completed. The assembly and testing structure was called the technical position.

Horizontal technical testing was carried out there. Several ground facilities also had to be built: photo-theodolite stations (for photographing the rockets during launch and flight), and a large weather station (to determine optimal weather conditions for launch). The basis of the working organization was a standard time system. A forward airfield was set up on a grass strip.

Completion of the test stand was important. Experience from Peenemünde and Lehesten was used in its design. It consisted of three levels and was equipped with the necessary launch and propellant tank equipment.

The horizontal testing was followed by the first burn test, but the engine could not be ignited. The three days of troubleshooting revealed a nonfunctioning relay in the launch electrics. Everything took place under the supervision of the first state commission chaired by Marshall of Artillery Yakovlev, Minister Ustinov, other ministers and generals, and Serov, Beria's deputy. The "Sword of Damocles" hung over them. One night the engine was finally ignited from an armored car that was serving as command post. It was the first ignition of a liquid-propellant engine at the central state rocket range.

An A-4 on the test stand during preparations for a burn test at Kapustin Yar.

View of the open tail section of an A-4, at the museum at the Luftwaffe Base at Cosford. Private collection of A. Kopsch.

Now everyone concentrated on the launch. The military launch crew consisted of members of the special purpose brigade, which in 1946, had worked almost the entire year at the Rabe and Nordhausen Institutes and was thus very well prepared. It was led by Engineering Maj. J.I. Tregub. The rocket industry's launch crew, also well-prepared rocket experts, was led by L.A. Voskrezenskiy.

The launch itself was carried out by the "Firing Section"; however, the date of October 18, 1947, on the plinth of the monument refers to the first launch by an A-4 (T) rocket built and assembled in Podlicki.

As the concrete command post at the launch pad had not been built yet, the launch took place from the captured German armored car.

At 0947 Moscow time on October 18, 1947, the first A-4 Series T rocket was launched. It flew 128.4 miles (170 miles according to other sources), maximum altitude was fifty-three miles, deviation to the left almost twenty miles. The impact crater

The two-part technical position and assembly hall.

Raising an A-4 rocket into firing position.

Raising a rocket on the launch platform, Kapustin Yar missile range 1947.

was not very large, as the rocket was destroyed when it entered the earth's atmosphere.

Yelena Hechayeva described the event as follows:

"Time is relentless. For sixty-five years many names and nicknames were extinguished from my memory. It is unlikely that any are still alive, but their work, their selfless self-giving in the name of the power of the Motherland can be heard in the roar of every missile taking to the skies. The lines left behind on the pages yellowed by time live again, written by the true and direct participants in the first launch of A-4 rocket No. 010 T by the commander of the servicing crew, Lt. Georg Vasilyevich Dyadin:

'Main (engine),' orders Korolev.

'Main (engine) on,' says operator N.N. Smirnitsky and presses the launch button on the console.

"And then the whole rocket is shrouded in smoke and flames, only the nose is visible. We watch as, staggering, it slowly begins rising from the raging sea of fire and smoke . . .

'At last!,' Gen. V.I. Vozynuk cries out.

"The rumbling reaches the roof. It shakes the earth and we can see the rocket heading upwards with increasing speed.

"A-4 rocket No. 010 T reached its calculated target area at 1007 Moscow time."

Above: Kapustin Yar missile test range, commemorative plaque honoring participants in the first launch of a long-range ballistic missile from the territory of the Soviet Union on October 18, 1947.
Below: Monument at the site of the launch of the R-1 ballistic missile, Kapustin Yar missile test range.

Raising an R-1 missile into launch position, 1948.

The missile has reached the vertical position.

The second launch, also of a T-Series rocket, took place on October 20. After launch the rocket veered sharply off course to the left. No news was received from the calculated impact area. Hours later, it was determined that the rocket had flown 145 miles and had veered 112 miles off course to the left, placing the city of Saratov within range.

Ustinov decided to bring in the German rocket experts Dr. Hoch and Dr. Magnus, who were present. "It's your rocket, your equipment, you look into it!" The results of the laboratory investigation revealed that certain vibrations produced an interfering signal that suppressed the wanted signal. Dr. Hoch developed a filter that suppressed the interfering signal. The next rocket to be launched deviated only slightly off course.

Ustinov's joy was boundless. He gave each specialist and assistant a bonus of 15,000 rubles and a canister of alcohol for everyone. The standing of the German rocket experts in the central commission was raised considerably. The series of rocket launches ended on November 13.

Overall, the results of the firing period were satisfactory. Of the eleven German rockets launched, five reached their targets. Five of the rockets had been assembled in Kleinbodungen and six in the NII-88 factory in Podlipki. Reliability figures matched the results achieved by the Germans during the war. All aggregates and other parts were German.

The launching of the A-4 rockets in autumn 1947, crowned the activity of the Soviet and German rocket experts in Germany. The tests had shown that the Soviet rocket experts had mastered the problems of practical replenished technology. They had amassed experience that opened the possibility of carrying on with the development of rocketry independently.

Launch position and launch of the A-4 rocket.

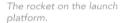

The rocket on the launch platform.

Delivery of liquid oxygen to the launch site.

Fueling the A-4.

The faults that had been discovered had to be rectified. A new solution had to be found for every problem, which proved very helpful in the development of the first Russian long-range missile, the R-1.

Cooperation with the German rocket experts and the exploitation of German experience led to enormous savings in financial resources and development time. Further valuable experience was gained by working on the central rocket testing site.

At the launch site the various groups worked together towards a single objective—the rockets must fly. The people and organizations got to know and understand one another better.

The participating ministers and their personnel, the military leaders and their staffs, and the scientists gradually came to understand that the rocket was not some sort of advanced projectile, rather a new and complicated system. There were no main and secondary tasks. All interests had to be subordinated to the final goal.

Working together in the close quarters of the launch complex made it possible for personnel to get to know the problems and possibilities even in other areas of expertise and promoted mutual understanding. Friendships were begun that lasted for years.

The Council of Chief Designers firmed up its cooperation. Its authority grew tremendously, which was of great importance to the work in the coming years.

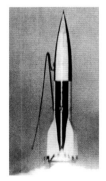

GORODOMLYA

The trip ended at some point—after twenty days and 1,200 miles. It was a train ride that left its marks. After five days a group under Gröttrup drafted and signed a formal protest. But not everyone shared their view. The food was good. Looking out the windows revealed untilled fields, shell craters overgrown with grass, wrecked tanks and aircraft, lone brick chimneys, few people. There was an initial stop in Moscow—a factory in the suburb of Podlipki. There was more waiting, then groups were formed and disappeared. At some point the rest learned that the train was heading for an island.

After more endless waiting the train began moving again. Another night journey. In the distance, a city could be seen on the shore of a big lake, surrounded by forests in autumn colors.

Ostashkov was a small city with a population of about 20,000. It was a typical small Russian city with, for the most part, unpaved streets. The final destination was the harbor. At the pier was a steamer with a tug. After half a day all of the furniture, boxes, and personal goods were stowed on the tugboat. The rocket experts and their families boarded the steamer. The vessel slowly made its way across Seliger Lake.

As it became darker the contours of the land disappeared. "a slightly cloudy sky reflected on the still water. This big lake was broad and still. Soon all its banks had shrunken into a thin line."

In the evening the families moved into their quarters. About the lake and the island they learned the following: it was in the area of the Waldai Heights, one hundred square miles. The island was about 2.5 miles long and up to 1.2 miles wide. The eastern part contained an inland lake. In the center of the island there was a group of houses, extending from north to south. As it was later learned, the island was surrounded by barbed wire.

As the island had been a test site for pathogens—hoof and mouth disease, leprosy, and the plague—the barbed wire probably originated from that time.

The most important work in the days that followed was setting up the apartments and laying in wood for the winter. Professional work was out of the question. It was put off.

By winter two single-story wooden houses were ready as workplaces. The scientific departments worked in one, the designers in the other. The promised materials, "paper and pencils," were on hand; however, writing paper had to be trimmed from rolls of packing paper. There was nothing else.

Financial compensation was based on qualifications, academic standing, and knowledge. Dr. Magnus, Dr. Umpfenbach and Dr. Schmidt received 6,000 rubles per month; Gröttrup and Schwarz 4,500 rubles; those with engineering degrees on average 4,000 rubles. Compare this to the pay received by

The city of Ostashkov (2010) on the south shore of Lake Seliger.

Lake Seliger.

The island of Gorodomlya.

the rocket experts of NII-88: chief designer Korolev 6,000 rubles; the institute's chief designer Pobedonostsev 5,000 rubles; deputy Mishkin 2,500 rubles; Chertok 3,000 rubles. The figures cited in various sources vary widely.

On days off and holidays it was possible for the Germans to visit the cities of Ostashkov and Moscow, and they could shop at markets and in stores and visit museums and theaters. The rocket experts and their families thus lived under very different conditions than German prisoners of war.

Soviet escorts always accompanied the Germans on these visits, as a rule acting as interpreters. They were partners in conversation for all questions, but also protectors against the inquisitive and pickpockets.

In winter, the children in particular, enjoyed being driven in sleighs across the ice of Seliger Lake to Ostashkov. In 1946, a school was established for children of all age classes. A blockhouse was made available for this purpose. German scientists and engineers provided the teachers. Gradually the curriculum was made to conform to that of Russian schools. The children first learned German with Russian as a second language. Later this was reversed.

In late 1946/early 1947, the tasks for the German rocket experts were laid down. The Russian documentation about the A-4 begun in Bleicherode was to be completed. A laboratory for the A-4 and guided missiles was to be established. The engine of the A-4 was to be examined and its thrust increased to 220,500 pounds. As well, the rockets completed in Germany were to be assembled.

The key point, however, remained the problems associated with the launch. How could they obtain the maximum amount of information about the rocket and its flight characteristics in a few flights (ten to twelve)?

In mid-1947, it was clear that the German teams were not capable of developing a new rocket complex. At Gröttrup's suggestion, however, the Germans were given the opportunity to develop a rocket project called G-1 (also R-4 or R-10). Gröttrup now had the title of chief designer. His deputy on the island was Dr. Wolff (ballistics). The organization on the island of Gorodomlya became "Branch No.1" of NII-88.

Another group of German rocket experts worked in the Podlipki facility until their relocation to Gorodomlya took place from 1947, until mid-

1948. They gained their own experience with Soviet industry and planning. The ritual had to be followed to realize a project: draft design – technical design – workshop project – experimental project – pre-production series. The scientists also found it difficult to get used to ending work when their shift was over. They were used to working on their problems well into the night. They worked alongside young, inexperienced engineers, some of them still in training. Working with the experienced German rocket experts would certainly have benefited them. After some time, it became obvious what was going on. A group of first-class rocket experts worked separate from the German rocket experts and studied their reports closely and also exercised a supervisory function. They gave questions to the young engineers and the Germans then had to answer them. It was an intentionally one-sided relationship. The Soviets were trying to obtain the knowledge of the German rocket experts without revealing their own. Contact with Soviet scientists in other fields was also undesirable.

Scientific work on the island strictly followed the rules of secrecy. Notes, files, books, writing and drawing paper, including slide rules and pencils, were passed out at 0800 and after lunch break. The reverse process took place at noon and at the end of shift. There was thus no lack of jibes and quips from the Germans. Secrecy requirements complicated their work. At every opportunity they were reminded not to reveal information to those not involved in the project. Pobedonostsev, by then the institute's chief engineer, put it to Chertok this way: "Don't you understand that our ruling bodies do not want the Germans to do actual work? They are under two levels of control. Under ours (as rocket experts) and under the organs of the NKVD, which in each of them see a fascist working for American intelligence. As well, whatever they produce would not be in keeping with our present ideological trend, as we have to regard everything in science and technology, new and old, as having been created without any contribution by foreigners."

Gen. Gonor, the director of NII-88, one of the first Heroes of Socialist Labor, was released from work because of his Jewish origins and then arrested. The struggle against cosmopolitanism developed. Every invention, discovery, and new scientific theory, without exception, was attributed to a Russian author.

Pleasant exceptions were the heads of the defense industry: Ustinov, Malyshev, Ryabikov, Kalmykov, Vetoshkin, and others.

Korolev also maintained no close relationships with the Germans. After the years of humiliation he wanted to finally design his own rockets. "You will not design your own rocket, you will redesign a German one," was the order. This went against the grain of this imperious, ambitious, and easily wounded man. Ustinov demanded production of the A-4/R-1 rocket as a school for Soviet industry.

The decision to exactly reproduce the A-4 rocket was dictated by the following considerations:

First, it was necessary to quickly form, educate, and train a large collective of engineers. It was also necessary to immediately give them a concrete and clear job to do and not to put it off into a distant future.

Secondly, the branch operations could not be left without work, lest they be taken over by some other contractor. Highly qualitative work documentation was, however, necessary to organize production. Where should they get it? They had had to develop their own from zero. Was it better to rework the German documentation? The answer was obvious, the second way was two years shorter.

Third, the military had already formed special units. As well, the State Central Rocket Testing and Evaluation Site was almost finished, and these could not be left with nothing to do!

Summer on the island of Gorodomlya.

Fourth, Soviet industry had to be provided with rocket technology as quickly as possible. Production of engines, equipment, valves, tubes, and plug-in connectors, for which the technical conditions and own blueprints already existed, had to be started as soon as possible.

And as this new cooperation was organized and developed on the concrete example of mass production of the R-1 rocket, Soviet rocket technology could make a leap forward and immediately proceed with development of its own rockets that could be used by the army.

At the same time, in 1947, the German collective was given the task of developing a ballistic missile with a range of 360 miles. Conflicts with Korolev were inevitable. There was also the fact that almost all scientific research departments worked for both Korolev and Gröttrup. Preliminary theoretical work on a missile with a range of 360 miles had begun at the Nordhausen Institute. Among those involved were Tyulin, Mishkin, Lavrov, Budnik, and others, who were now working under Korolev.

THE R-1 AND G-1 MISSILES

In February 1947, Korolev said:

"It would be a mistake to assume that the creation of the Soviet R-1 rocket was simply a matter of copying German technology, only with Russian replacement materials. Apart from the change of materials and the redesign of the entire technical manufacturing process for the components and parts of the rocket, one must consider that the Germans never got the A-4 rocket to a state necessary for use as a weapon. The experience of studying German rocket technology shows that the Germans committed a great deal of effort and resources to completing this task, meaning final development of the A-4 rocket. Simultaneous with the research and design work, in many installations the Germans worked on scientific concepts, both of an operational and a perspective character.

"Until now we have not made any flight experiments with rockets captured in Germany, and consequently we have no definitive expectations of this design.

"Many of these and other questions will have to be investigated on a large scale in our research and development, in our scientific-technical institutions, institutes, companies, test stands, and at the launch sites during the development, production, and manufacture of the first batch of Soviet R-1 rockets. To achieve this, flight experiments with the A-4 rockets that have been in storage at the NII for a long time are necessary. In this way we will gain the necessary practical experience and can assign new tasks to our work collectives in the field of long-range rockets. Now it is necessary to begin equipping a site and laying out the routes on the firing range for conducting flight tests and also to begin construction of an experimental stand."

In 1947, Korolev's department was hard at work on the R-1 rocket, but at the same time also on the R-2 with a design range of 360 miles. Korolev tried to fully integrate the experience gained with the A-4 into the R-1 rocket.

The diameter of the R-1 was identical to that of the A-4. This made it possible to install the more powerful engine, which was delivered by Design Bureau 456 under Glushko. The government confirmed the plan of development for a rocket with a range of 360 miles. Also envisaged was the R-3 rocket with a range of 1,800 miles. Korolev

By Lake Seliger.

foresaw the difficulties that such a big step would bring with it. Ultimately everything depended on the engine, meaning everything depended on Glushko and his OKB.

Here the experience gathered during engine trials at Lehesten was used to increase engine thrust from 55,100 to 77,200 pounds. This was sufficient to propel an A-4 carrying a 2,200-pound payload a distance of 360 miles. At the same time, of course, the weight of the rocket had to be reduced. One possibility consisted of allowing the propellant tanks to become part of the outer shell by increasing internal pressure. Another issue was identified in early 1947. Why did the entire rocket have to fly to the target? Was not the nosecone with the explosives sufficient? This would remove one of the problems, that of the rocket body's structural strength, from the table. At Peenemünde this type of breakup was called "air burster," however it was never investigated further.

The G-1 was the first design by the German rocket experts under Gröttrup's leadership. A group of 234 rocket experts was given the task of designing a rocket with a range of 360 miles. The work began in Germany. The first discovery in the Soviet Union was that the technical documents were still in transit, somewhere between the central works and Podlipki.

During development of the G-1 by the German rocket experts another question became more important: how hot would the outer surface of the rocket become during flight? Dr. Albring calculated the temperature taking into consideration that the nose of the rocket would compress the air and over time considered altitude and velocity. However, the temperature that had been measured automatically at Peenemünde was considerably higher than that calculated by Dr. Albring. But this temperature was of vital importance to the rocket's structural integrity. Further calculations then resulted in an increase to 3,632 degrees Fahrenheit.

Dr. Werner Albring, aerodynamicist.

What material could withstand that? As well, there was also friction over the entire outer surface of the rocket. The calculated temperature was that of the surrounding air, but how long would it be before the outer surface of the rocket became heated? All of the available data was based on experience in aircraft construction, but the flight path of a rocket was very different. However, with increasing speed (passing through the sound barrier) and altitude the density of the air diminished. Not until the descent phase of flight did speed and air density and thus temperature begin to rise. This explained the instability of the missile and why the A-4s had broken up in the air. But the future G-1 rocket was to fly even faster and further. Was this the end?

Gröttrup, falling back on an old idea from Peenemünde, had the solution. Separate the forward part of the rocket with the explosive charge after the burnt-out missile passed the apex of its flight path. The warhead was mechanically reinforced and fell to the target. The rest burned up in the atmosphere. Thus the German and the Soviet rocket experts independently came to the same realization. Separation of the nosecone was tested with a modified version of the R-1, called the R-1A. Six launches were made between May 7 and 28, 1949, and were successful.

Another hindrance was the Russian production technology, which was equivalent to that of Germany at the beginning of the 1930s. The German rocket experts worked at two sites, in NII-88 (Korolev's OKB) and on the island of Gorodomlya, to complete the design of the G-1. Other groups of Germans worked in OKB-88 (R-1 production) and OKB-456 (Glushko/engine production).

In September 1947, the German rocket experts presented their G-1 project to the scientific-technical council of NII-88.

The meeting was chaired by General Gonor, head of the NII. Also taking part in the discussions were: Vetoshkin, head of the Main Administration for Rocket Technology in the Ministry of Armaments; Pobedonostsev, the institute's chief engineer; Tikhonravov, pioneer of rocket technology; chief designers Ryazanski, Pilyugin, and Kuznetsov; Nikolayev, director of the Baumann Technical University in Moscow; Chertok, the chief designer's deputy; Trapeznikov, director of the Institute for Automation at the Academy of Science; Professor Kozmodemyansky; and deputies Mishkin and Busheev. Korolev also attended the meeting of the scientific-technical council.

The main presentation was made by Gröttrup, who was in charge of the design project. There to support him were several German rocket experts: Dr. Umpfenbach and Doctors Hoch, Albring, Anders, Wolf, and Schefer.

In his presentation Gröttrup listed the following advantages of his missile: it retained the external dimensions of the A-4. The propellant tanks were an integral part of the outer shell. This resulted in an increased quantity of propellant and doubled range. The control system was simplified by transferring control functions to ground radio stations, resulting in a reduction in the number of electronic devices, plug connections, and wiring. The result was a simplification of the rocket and a tenfold improvement in accuracy. Separation of the warhead took place during the unpowered part of the flight path. The time required for launch preparations could be halved.

An assessment of the rocket's effectiveness brought the presentation to an end. The destruction

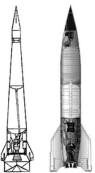

Drawing and cutaway of the G-1 missile.

of an area .93 x .93 miles in size at a range of 180 miles would require the use of 67,500 V-2 rockets, but just 385 G-1 missiles would be necessary to destroy the same area at a range of 360 miles.

A basically new method used in designing the control system was the use of flight path models. This saw the first use in the Soviet Union of an electro-mechanical analogue modeling machine. It was possible to model equations of motion with respect to the rocket's center of mass by using variable coefficients, and to solve these equations taking into consideration the parameters of individual devices that were fed into the models. Thus they could check out the equipment of the A-4 rocket prior to launch, explained the author of the model Dr. Hoch. Dr. Hoch was one of the talented German scientists. Together with Dr. Magnus he developed the totalizing gyroscope. He died a short time later from suppurative appendicitis.

The following are extracts from the ruling of the scientific-technical council:

"The presentation on the G-1 missile contained a number of essentially new solutions to individual design elements of the missile. Overall the project must be assessed positively. Particularly interesting is the missile project's control system, which solves the question of improving the A-4 rocket's accuracy. The presentation and the subsequent discussions revealed, however, that many significant elements of the control system are not yet developed and do not meet the requirements to be placed on the project . . .

"Separation of the warhead from the missile is new and deserving of positive assessment . . .

"Load-bearing propellant tanks made of light alloy can result in considerable lightening of the design in the center section of the G-1 missile compared to the A-4 . . .

"The G-1 (R-10) engine project makes it possible to simplify the entire engine structure and thus reduce the outer dimensions of the engine. The practicality of driving the turbine with gases drawn directly from the combustion chamber will undoubtedly have to be confirmed through experimentation.

"It is necessary to speed up the detailed development of the overall control system. The most important components are to be modeled.

"The radio portion of the project must be assessed by qualified experts.

"The project will once again be considered as a missile project at one of the following regular meetings of the plenum of the scientific council."

Formally, Gröttrup and his collective could not protest the decision by the Scientific-Technical Council. Actually, not only was the Scientific-Technical Council in a very difficult situation, but also the head of the institute and the Ministry of Armaments, at whose direction this project was carried out.

Later improvements would increase range to 503 miles with a target accuracy ellipse of 0.75 x 0.80 miles. The G-1 was given the secret designation R-4 and the overt designation R-10. Also discussed quite actively was the G-1M (R-13) missile project with the body of the G-1 and a more powerful propulsion system from the V-2.

The simultaneous realization of Korolev's plans was not possible for engineering-technical and production reasons, because the necessary work forces were simply not available.

The German rocket experts had passed Korolev. His duties left him no time: preparing the A-4 launches at Kapustin Yar, organizing production of the R-1, and organizing his work at NII-88.

On April 14, 1948, the government passed the order for the development of the first missiles from domestic materials, based on the German A-4 rockets. These missiles, together with their complex of ground equipment, were called R-1.

The R-1 missile was obsolete before it was born, as the R-2 missile project, which had much better flight and tactical characteristics, was ready. To a certain degree, however, the decision to complete work on the R-1 and its adoption by the Soviet Army was necessary.

The experience gained with the A-4 rockets made it possible for the Soviets to concentrate their full attention on organizing production. It was a task of vital importance, as there was no branch of Soviet industry suited to produce the missiles. Thirteen research institutes and thirty-six companies had taken part in development of the R-1. The R-1 missile was developed by organizations under the leadership of S.P. Korolev (missile, launch complex), V.R. Glushko (engine), N.A. Pilyugin (control system and ground control and launch apparatus), V.R. Barmin (ground launch, propellant tank, and other equipment), and V.I. Kuznetsov (guidance system).

As a result of the shortcomings of the A-4 rocket demonstrated during flight testing, and also the almost complete absence of theoretical documentation with justifications of the technical decisions that had been made, development of the R-1 missile required as much work as a new design. Material-technology tasks proved particularly labor-intensive.

The material specialists could not be permitted to limit the formal selection of domestic materials. They demanded a critical analysis of the technical decisions taken from the Germans. Required were eighty-six types of steel, fifty-nine types of non-ferrous metal, 159 types of non-metallic material, and the like. Domestic industry only produced thirty-two types of steel and twenty-one of non-ferrous metals. Most difficult was the replacement of rubber, inserts, insulation, plastic, etc. These materials were required to improve reliability of the missiles.

During A-4 rocket launch preparations, Soviet specialists had observed numerous cases of unsealing*, faults caused by poor quality of materials and coverings, and incomplete technical and design work. Therefore during the design of the R-1 the problem of ensuring the long-lasting integrity of its design had to be solved. This included not just the creation of effective protective coverings, but also an essentially new approach to the selection of non-metallic materials with expanded sources, plus the development of methods of protecting metal structures against corrosion and investigating the corrosion resistance of the design.

Cutaway drawing of the short-range R-1 missile, also designated 8A11, SS-1 Scunner, R-1M, and R-1A.

* When an air-tight (hermetically sealed) container loses air as the result of a leak, this process is called unsealing.

Despite the obvious possibilities, copying of the A-4 rocket for the first series of R-1 missiles was limited, as the engineers immediately strove to introduce their own technical solutions. But the tight schedule did not allow this: the tail and equipment sections were to be redesigned to improve structural strength. And computed range was to be increased to 150 to 162 miles, taking into consideration the increased amount of propellant (ethyl alcohol) on board.

The engine of the A-4 rocket was used in development of the power plant for the first series of R-1 missiles, without any design changes, with the exception of the replacement of much of the

materials with domestic products. Some rubber-metal parts had to be taken from captured stocks.

The wiring diagram of the A-4 rocket was also adopted unchanged for the first series of R-1 missiles;

Cosmonaut Valeri Nikolayevich Kubasov in front of a V-2 engine in Bleicherode, 2010.

RD-100 (8D51) liquid-fuel rocket engine that powered the first R-1 (8 11) ballistic missile, which entered the Soviet inventory in 1950. Chief designer V.I. Glushko, OKB-456.

*View into an open V-2 engine at the Luftwaffe Base
Museum, Cosford. Private collection A. Kopsch*

however, the design and technical characteristics of many devices had undergone changes.

Based on the Messina telemetry system for the A-4 rocket, the new Brazilianite tracking system with an increased number of channels was developed for the R-1.

The R-1 missile of the first production series had a maximum range of 168 miles, a length of 46.75 feet, maximum diameter of 5.4 feet, and diameter with stabilizing fins of 11.7 feet. The missile's launch weight was 28,750 pounds, while the missile weighed 8,818 pounds empty. The propellant components (liquid oxygen and 75% ethyl alcohol) weighed 20,822 pounds and the warhead 1,800 pounds. Maximum height of the flight path was forty-eight to fifty-one miles, engine thrust on the ground 55,115 pounds. Specific impulse was 656 feet per second, velocity at engine burnout 4,593 to 4,921 feet per second, maximum deviation in azimuth and elevation twelve miles, and engine burn duration sixty-five seconds.

As the rocket was prepared for production, work on the complex of launch equipment at the provisional complex went ahead (concreting, lay-ing of foundation stones beneath the launch pad, underground shelters for the movable diesel stations and other equipment, and the laying of underground cables).

The technological processes involved in the equipment used to prepare the R-1 missile for launch, and the design of the assemblies scarcely differed from those of the A-4 rocket. V.R. Barmin was named chief designer for the R-1 missile ground complex. The State Special Design Bureau (GSKB) Spez'masch, the most important organization for missile ground complexes, was later created under his leadership.

The main parts of the missile were: the nose section, the equipment compartment, the propellant tank, the oxidizer tank, the tail section with engine. The main feature of the rocket's design was the use of a non-separating nose section with the use of suspended (non-integral) propellant tanks, which were mounted in the main body. The main body of the rocket consisted of a rigid skeleton of steel beams and formers with a shell of sheet steel. The propellant and oxidizer tanks were fabricated from aluminum alloy sheet.

Four large and heavy fins (weighing approxi-mately 660 pounds) were required to stabilize the

Specifications for the A-4 and R-1 Rockets		
Designation	A-4	SS-1A; 8K 11; Scunner; 8A11
End of Service	1952	1964
Rocket Launch Weight	28,230 lb	29,206 lb
Payload	2,200 lb	1,796 lb
Empty Weight	8,836 lb	8,818 lb
Rocket Length	39.3 ft	46.75 ft
Maximum Diameter	5.4 ft	5.4 ft
Wingspan	11.67 ft	12 ft
Thrust	70,095 lb	61,035 lb
Specific Impulse in a Vacuum	239 kpa	200 kpa
Specific Impulse on the Ground	203 kpa	200 kpa
Burn Duration	69 sec	65 sec
Range	max. 170 mi	205 mi
First Launch	Oct. 18, 1947	Sept. 17, 1948
Last Launch	Nov. 13, 1947	Sept. 13, 1964
Number	56	114

Left: Engine test stand.

Right: Transport of an R-1 missile.

missile in flight. Two types of control mechanism were required: air rudders (on the stabilizing fins) and jet rudders (in the exhaust stream from the engine nozzle).

The single-chamber liquid-fueled rocket engine operated on a propellant of liquid oxygen and 75% ethyl alcohol. The propellant delivery system consisted of unsealed pumps (the gas used in the turbine was vented into the atmosphere). Steam formed by the decomposition of hydrogen peroxide in the presence of a catalyst (a solution of sodium permanganate) produced steam, which was used as the turbine operating element. The delivery of peroxide and permanganate into the reactor had a displacement effect. In this way four liquid components were needed for the engine to operate. Their consumption rates were: 165 pounds per second of liquid oxygen, 110 pounds per second of ethyl alcohol, and 3.75 pounds per second of peroxide and natrium-permanganate. Specific impulse was 6,630 feet per second on the ground and 7,762 feet per second in a vacuum.

Such low values for the specific impulse are explained by the use of a low-energy propellant (the propellant was supplemented with water, as there was no other way to cool the chamber), the low parameters of the engine's operating process, and the use of the open engine system scheme.

The engine was heavy, which was explained by the incompleteness of the design of all its main components: the combustion chamber (the low

pressure, the poor organization of the propellant combustion process), the tube pump assembly (the low number of revolutions), and the steam generator (components delivered by a displacement system). During engine start, the propellant in the combustion chamber was ignited by a pyrotechnic igniter. The ratios that determined speed and therefore range, were extremely low on the R-1.

The rocket employed an autonomic inertial guidance system, which used the shape of the rocket's angular position during the powered phase of flight and the automatic system for lateral control, in which the accelerometer-gyro integrator was used. The control system had a significant weight (weight of the control devices was approximately 440 pounds, with the basic weight of the equipment section being 1,150 pounds). Overall, the rocket's accuracy (0.93 miles) can be rated as poor, considering that it is combined with a range of only about 180 miles.

Warhead size depended on the target, and approximately 1,750 pounds of explosives could be accommodated in the nose section. The area of destruction in a city was therefore no more than sixty-five to eighty feet. The rocket could therefore only be used for area bombardment of large, weakly-protected targets.

The technical ground equipment associated with the complex included more than twenty special machines and pieces of equipment. A crew of eleven, at the technical and launch positions, prepared the

Transporting liquid oxygen to the launch position, 1948.

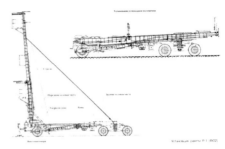

Cutaway drawing of the 8U22 lowered carriage in travel and upright positions.

rocket for launch. The principal tasks at the technical position were checking the rocket's systems and attaching the nose section.

Rocket check-out at the launch position was carried out with the 8U22 or 8U24 lowering carriage, with whose help the rocket was then placed on the launch platform. It was also used in preparing the rocket for launch.

After the rocket had been placed in a vertical position, its control system was checked out. The tanks that held the propellant and steam-producing materials were filled. Aiming was carried out. The operating team had a special responsibility while tightening the removable, rotatable operating position on the rocket's warhead. Careless execution of this operation could result in the support ring coming into contact with the spot for the warhead fuse, which was at the tip of the rocket.

Manual operations with the rocket engine were also carried out in preparing the rocket for launch—tuning the steam generator's pressure-reducing drive as a function of hydrogen peroxide concentration and temperature. With these engine parameters they approached the nominal. The ignition device was installed in the engine's combustion chamber through the nozzle. The rocket was fired from a special armored car with a console. The time to prepare the rocket at the technical position was two to four hours, and four hours at the launch position. The time required for the complex to reach combat readiness, meaning the time from receiving the order to launch until the launch of the rocket itself, was no less than six to eight hours. After that a decision could be made to launch the rocket or delay the launch until the next day. Filling the oxygen and propellant tanks and checking the system took the most time.

Left: Raising the rocket on the Meillerwagen. Right: Raising the rocket with the aid of a support frame.

The rocket on the launch platform.

R-1 missile in launch position, 1948.

Technical Position

In rocketry—the complex of structures and technological tools envisaged for guaranteeing the preparations and the launch of the rocket. The spaceflight center's technical position consists of the complex of structures with general technical and specialized technical equipment and access roads for the acceptance, storage, assembly, testing, and fueling of cosmic rocket technology. The main structures of the space center's technical position are the assembly-inspection building, the fuel storage site, and the fueling location. The technical position can be mobile or stationary. Its equipment can now include about a dozen machines and installations, which all guarantee work during use of the rockets.

Launch Position

The part of the grounds on which one or several launch sites, equipment, and special structures are set up, which are predetermined for the contents of the rockets at a certain stage of combat readiness, their technical servicing, the pre-launch preparation of the launch. The launch positions can be main or secondary positions, open or closed, individual, and in groups. (*Source*: http://military_terms.academic.ru)

Names of Position Elements

- Бронемашина управления
Armored car for launch control
- Пусковой стол с ракетой
Launch platform with rocket
- Подогреватель-заправщик перекиси волорода
Hydrogen-peroxide servicing tanker-heater
- Автоцистерна-заправщик спирта
Alcohol servicing truck
- Кислородная цистерна
Oxygen tank
- Электропреобразовательный агрегат
Electric power converting unit
- Бензоэлектрические агрегаты
Gasoline-engine electric sets

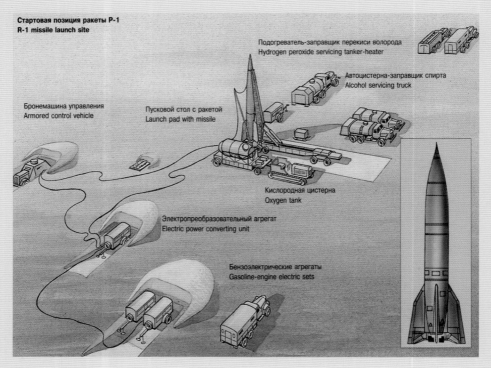

Стартовая позиция ракеты Р-1
R-1 missile launch site

Подогреватель-заправщик перекиси волорода
Hydrogen peroxide servicing tanker-heater

Автоцистерна-заправщик спирта
Alcohol servicing truck

Бронемашина управления
Armored control vehicle

Пусковой стол с ракетой
Launch pad with missile

Кислородная цистерна
Oxygen tank

Электропреобразовательный агрегат
Electric power converting unit

Бензоэлектрические агрегаты
Gasoline-engine electric sets

R-1 missile launch position.

If one overlooks the obvious design shortcomings, the R-1 was an almost-exact copy of the German V-2. Nevertheless, the R-1 rocket played its historic role. This allowed the USSR to create in a short time all the conditions needed for the further development of the new class of weapon and decided the path and direction of this development. Before 1946 was over, meaning before work on the R-1 began, the first rocket unit of the Soviet Army was formed—the Special Purpose Brigade of the High Command Reserve.

Cutaway drawing of the R-1 missile.

The brigade's personnel familiarized themselves with the new weapon while still in Germany. Then it took part in launches of A-4/V-2 and R-1 rockets at Kapustin Yar. Based on the experience it had gained, the Special Purpose Brigade began working on documents for personnel training and the operational use of long-range rockets.

Flight testing of the first series of R-1 rockets with the complex of ground equipment was carried out at the central rocket range in September-November 1948.

The first launch was made on September 17, 1948; after launch the rocket deviated fifty-one degrees off course and crashed. The suspected cause was failure of the guidance system. The rocket had been built by Research Institute 88's facility in Podlipki.

S.N. Vetoshkin was named chairman of the state commission during flight testing of the R-1 rocket, and S.P. Korolev was named technical director. V.P. Glushko, V.R. Barmin, V.I. Kuznetsov, and N.A. Pilyugin, all members of the Council of Chief Designers, were added to the technical leadership.

Nine R-1 rockets were tested in the first phase. The first launch took place on September 17, 1948. The results of these trials were extremely disappointing: of nine rockets, just one reached its target (first successful launch on October 10, 1947, other sources say October 31). The causes of the problems were very different, but mainly of a technological nature: poor quality in the manufacturing of rocket assemblies and systems, insufficient testing of junctions and equipment, poor workman-

ship in some systems. At the same time, major difficulties developed during the tests. During the launches strong blows were heard in the engine. After the first series of test launches, it was found that the blows had been caused by the pyrotechnic ignition devices in the engine. To rework the ignition devices of the R-1 rocket a stand was installed on the preliminary stage capable of engine starts. During the first test on the stand on April 30, 1949, the engine could not be shut down at the specified time. The preliminary stage ran for about nine minutes, until the fuel components were completely exhausted. The situation was dangerous. The engine on the stand could have exploded. Nevertheless, this occurrence served as cause for the more careful execution of the launch operation and the shutting down of the rocket's engine.

During subsequent engine tests a month and a half later, the powerful blows reappeared in the engine. Only the use of liquid-fuel ignition system, proposed by OKB-456 MPA (Ministry of the Aviation Industry) (V.R. Glushko) allowed the powerful blows during start-up of the R-1 rocket's engine to be eliminated. Liquid-fuel ignition systems were subsequently used in all rockets in which oxygen was used as the oxidizer. To the shortcomings of the A-4 were added the new faults resulting from Soviet development.

In spring 1949, the first issue of *Operational Rules for the Crews* of the R-1 Rocket was issued. These rules were tested during the second series of rocket launches.

The first test of the R-1A rocket—a modification of the R-1 for testing the separating nose section—

Tactical-Technical Characteristics of the R-1 (8A11) Rocket	
Maximum launch weight, lbs	29,541
Payload, lbs	1,796
Empty weight of rocket, lbs	8,818
Weight of warhead with conventional explosives, lbs	1,730
Weight of warhead, lbs	2,200
Weight of fuel, lbs	18,739
Maximum firing range, mi	167
Maximum velocity, ft/sec	4,806
Maximum altitude of flight path, mi	47.8
Approximate flying time, min	5
Engine thrust on ground/in a vacuum, lbs	59,493/68,308
Specific impulse of engine on ground/in a vacuum, kpa	199/232
Main engine operating time, sec	65
Target accuracy, mi	0.93
Type of nose section	Monoblock, non-nuclear, non-separable
Rocket length, ft	47.9
Maximum rocket diameter, ft	5.4
Engine type	RD-100
Main engine mass, lbs	1,951

was carried out on May 7, 1949. Four launches took place in the first test series. On the fifth and sixth launches the rockets carried a payload of scientific equipment.

To improve reliability, for the second series of tests a virtually new underground cable network had to be installed and many changes made to the on-board devices of the control system. The second series of rockets was equipped with the new Don telemetry system.

Twenty rockets were prepared for the second series of tests; ten for firing, and ten for assessment. Afterwards, rockets carrying scientific equipment were given the designations W-1B, W-1W, W-1D, and W-1E.

The R-1A Experimental Rocket

To perfect the weight and operating characteristics of the R-1 rocket, it was proposed that the nose cone be separated from the rocket after engine burnout. Theoretically, only the powered portion of the flight was left for the booster, which in terms of mechanical and thermal loads was much more favorable than the atmosphere during the descent phase of the flight path. The R-1A rocket was created to experimentally verify these new ideas, mainly to study behavior during separation of the nosecone. The separating nose section was first used in the design of the R-2 rocket, but because many organizations were interested in using the new rockets for their own purposes, the experimental program development far beyond the scale originally envisaged.

For development of the new separating nose section, the R-1A rocket's control system was equipped with special gyroscopic devices and telemetric measuring equipment. To obtain additional information, special optical systems were used to observe the launch (submarine periscopes, photo-theodolites, movie theodolites) for the purpose of observing the behavior of the nose section during the passive phase of the flight path. Radar was also envisaged for active tracking. For this purpose special transmitters were attached to the outer surfaces of the rocket. After separation of the nose section, the radio signal was transmitted to the telemetry system of the Physical Institute of the Academy of Science of the USSR (ФИАН-PIAW). Separation of the nose section proved such an effective solution that it was used in the design of all subsequent domestic and foreign rockets.

After the positive results of launches with ballistic flight paths, it was envisaged that two R-1A rockets would be launched vertically for physical research of the upper layers of the atmosphere with the aid of the ФИАН-1 system of the geophysical institute of the Academy of Science of the USSR. Here the idea of the separating container was born, and a new completely sealed ФИАН-1 container (PEAR = physical research of the atmosphere with rockets) was designed.

R-1A rockets were also used to conduct research into the effects of the engine gas stream on the passage of radio waves. This was directly linked to the development of radio control systems for the R-2 and R-3 rockets. It was planned to set the engine to a lower thrust level, so that the ratio of initial mass to the thrust of the engine would be the same as that of the R-2 rocket. The R-1A rockets underwent

R-1A Rocket – General Technical Information	
Suborbital ballistic short-range booster rocket	
Launch weight	29,541 lb
Height	46.25 ft
Diameter	5.4 ft
Range	62 miles (37 miles)
First launch	May 7, 1949
Last launch	May 28, 1949
Number	6

preflight checks at Research Institute 88's temporary testing site in January and February 1949.

Flight tests with the R-1A rocket were begun at the State Central Rocket Range in May 1949, and the results were positive. A total of six launches were carried out—four with ballistic flight paths and two vertical.

The launches were made with an increased engine thrust to mass ratio of 1.97 instead of the R-1 rocket's 1.83 to assist in calculating the engine system regime for the later rockets. The PIAV receiving station was deployed to observe the flight of nose section after separation in the area where it descended into the landing area.

The second series of launches took place between September 10, and October 23, 1949. During the tests in autumn 1949, seventeen of twenty rockets completed their missions. Additional experimental work was required to ensure trouble-free launches by the R-1 rocket.

Numerically, the results of the second series of flight tests with the R-1 rocket were satisfactory. Of twenty rockets, seventeen fell in a rectangle of ten by five miles, which was determined by tactical and technical requirements. Just three rockets failed to reach the target: one because of loud noises, whose shocks led to the premature unlocking of the integrator; and the others because of an error in tuning the integrator. Two rockets were damaged in the launch area: technical unsealing of engine communication as a result of loud noises, and the explosion of the oxygen tank during fueling, caused by a defect in the drainage valve resulting from pressure loss. There were no failures during launch as a result of postponements. Pilyugin and his people were very proud of this, although the loud noises seriously affected the nerves of the testers, as they had earlier.

Six of the ten first series rockets were removed from the launch platform because of failures during launch. None of the twenty second series rockets was removed.

The launches in September and October were carried out in a significantly more pleasant atmosphere. The rocket specialists lived in hotels and ate in restaurants. All of the roads were paved and the wooden hall for horizontal testing was replaced by a comfortable assembly and testing building.

After the launches were over, an editorial committee was formed. The committee worked "from dawn till dusk," until the stenographer-typists were exhausted. The conclusions and findings were changed and ten copies were printed. Myrkin calculated that the problems associated with the rockets were still so serious that it would be too early to begin production of a large series. It was not especially necessary to recommend its introduction into service. Korolev was extremely dissatisfied with such a position. He, for example, insisted on roughly formulated wording: "Let us begin mass production. We will overcome the problems that have been identified in the process of flight testing." These differences had to be decided in Moscow at the level of ministers and marshals. For Ustinov, Vetoshkin, Gonor, Korolev, and all others, the makers of the R-1, mass production began with the formulation "acceptance into the arsenal." After returning from the rocket range, stock was taken at an informal meeting. Korolev demanded that NII-88 bring order and culture to the companies. Mechanical work presented no problems. New technologies were only accepted with great reluctance; for example,

technologies for welding and riveting when working with non-ferrous metals and copper. This was necessary for self-assertion in the new technology, to increase the authority of entire production facilities.

At the end of 1949—after four years of hard work—the prestige of the rocket builders demanded that they be able to begin production of the R-1 rocket. The Germans had achieved this in 1944. OKB-456 spent almost all of 1950 experimenting with the new liquid ignition system, instead of the pyrotechnic one, to eliminate the pulses. Kuznetsov worked on the pulse resilience of the integrator. Pilyugin tormented his subcontractors with the objective of improving the reliability of all relays and contact connections. Chertok yet again declared to the production workers the necessity of cleanliness and work culture in the manufacturing of guidance systems. A capable helper soon appeared in this sector: Viktor Kalashnikov.

The R-1B, R-1W, R-1D, and R-1E Geophysical Rockets

The experiments with the FIAR-1 device for physical measurement of the thin upper atmosphere served as the basis for the preparation of a broad program of scientific research in the field of geophysics on behalf of the Academy of Science of the USSR and the development of modified versions for the R-1 rocket (R-1B, R-1V, R-1D, and R-1A), which were envisaged for this project. To coordinate the work, the Academy of Science of the USSR created a special committee chaired by S.I. Vasilov, president of the Academy, and his deputy M.V. Keldysh. The work was carried out in keeping with an order from the Council of Ministers of the USSR on December 30, 1949.

R-1B

All four launches of the R-1B rocket in July–August 1951, were vertical launches, the first of which failed. On board the rocket were research animals in a special sealed container. Their behavior in a

R-1B Missile: General Technical Information	
Short-Range Suborbital Ballistic Missile	
Launch Weight29,542 lb	
Height	57 ft
Diameter	5.4 ft
Range	62 mi (37 mi)
First Launch	July 29, 1951
Last Launch	September 3, 1951
Number	4

R-1W Missile: General Technical Information	
Short-Range Suborbital Ballistic Missile	
Launch Weight	29,542 lb
Height	57 ft
Diameter	5.4 ft
Thrust	60,025 lb
Range	62 mi (37 mi)
First Launch	July 22, 1951
Last Launch	August 19, 1951
Number	2

weightless environment was examined. The rocket came down by parachute so that it could be used again for further experimental launches.

The R-1B version was designed for scientific research at altitudes up to sixty-two miles, including the investigation of cosmic rays. It was also used to investigate the characteristics of the atmosphere, and the effects of zero gravity and radiation on animals.

R-1V

The R-1V rocket differed from the R-1B only in having a rescue parachute in the rocket body in place of the FIAN apparatus. A total of four launches were

made in July–August 1951, two of which failed.

The R-1V version was developed for scientific research at altitudes up to sixty-two miles, including the investigation of cosmic rays. It was also used to investigate the characteristics of the atmosphere, and the effects of zero gravity and radiation on animals. It was used to test the practice of recovering the rocket by parachute and reusing it for further experimental launches.

R-1D

Unlike the R-1B, in which research animals and their sealed capsule were recovered, the R-1D carried two dogs aloft, each in a pressure suit that was attached to a special cradle with a parachute and a life-support system. As well, instead of the compartment with the FIAN apparatus the R-1D had an apparatus for researching the distribution of ionization based on altitude, density in the ionosphere,

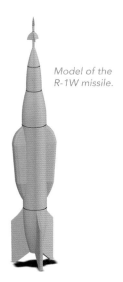

Model of the R-1W missile.

R-1D Rocket: General Technical Information	
Suborbital ballistic short-range booster rocket	
Launch weight	29,541 lb
Height	57.4 ft
Diameter	5.4 ft
Range	62 miles (37 miles)
First launch	June 26, 1954
Last launch	July 7, 1954
Number	3

and for the propagation of super-long waves in the atmosphere and in cosmic space. The launches of the R-1D were carried out in June–July 1951. All were successful.

R-1E

With the R-1E another attempt was made to find a design solution that was to ensure the recovery of the rocket body. For this objective three powder boosters were attached to the nose section, which gave it a separation velocity of about forty feet per second. This measure proved to be inadequate, however. The new variant of the recovery system involved the use of a pyro cannon. Not only was it supposed to lead release pilot chutes, but also

simultaneously free the parachute package, which was in the main part of the cone. A total of six launches were made between January 1955, and April 1956, of which four were successful.

The R-1E version was used for scientific research at altitudes up to sixty-two miles. It was used to investigate winds in the upper layers of the atmosphere, the composition of the air, solar radiation, the ionosphere, and characteristics and effects of spaceflight on living creatures, and their readjustment to terrestrial conditions.

Tests of the R-1 rockets used for scientific research lasted until May 1956. All of the tasks associated with the use of the R-1 rocket and its modifications for the conduct of scientific experiments were completed with one exception—recovery of the rocket body itself.

R-1E Rocket: General Technical Information	
Suborbital ballistic short-range booster rocket	
Launch weight	29,541 lb
Height	46.25 ft
Diameter	5.4 ft
Thrust	60,023 lbs
Range	62 miles (37 miles)
First launch	January 25, 1955
Last launch	May 31, 1956
Number	4

R-1E (A-1) Rocket: General Technical Information	
Suborbital ballistic short-range booster rocket	
Launch weight	29,541 lb
Height	46.25 ft
Diameter	5.4 ft
Range	62 miles (37 miles)
First launch	February 5, 1955
Last launch	June 7, 1956
Number	2

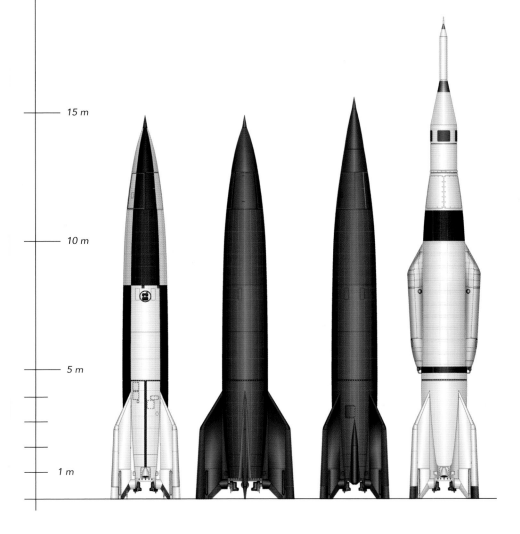

The R-1 and several modifications.
R-1 (8A11), R-1A (V-1A), and R-1E (V-1E).

R-1
First launch on
September 17, 1948.

R-1 (8A-11)
First launch on
August 20, 1950.

R-1A (V-1A)
First launch on the
FIAR-1 container on
May 24, 1949.

R-1E (V-1E)
First launch on the
FIAR-1 container
January 25, 1955.

15 m

10 m

5 m

1 m

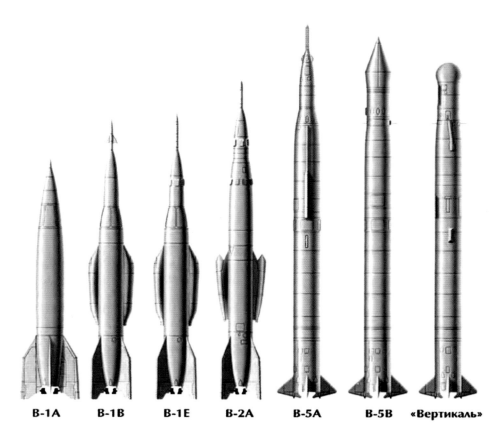

В-1А В-1В В-1Е В-2А В-5А В-5В «Вертикаль»

Illustration of the geophysical rockets based on the R-1, R-2, and R-5 missiles: V-1A, V-1V, V-1E, V-2A, V-5A, V-5V, and Vertical (Umanski, S.P.: Booster Rockets, Spaceflight Centers, 2001).

The letter B in the designation B-1A was derived from the Russian word Вертикаль—the Vertical.

В-1А **В-1Е**

Под редакцией:
Героя Социалистического Труда
академика Василия МИШИНА,
дважды Героя Советского Союза,
летчика-космонавта СССР
Владимира АКСЕНОВА
Коллективный консультант:
Государственный музей истории космо-
навтики имени К. Э. Циолковского.

ГЕОФИЗИЧЕСКИЕ РАКЕТЫ	В-1А	В-1Е
Стартовая масса, кг	13 910	14 211
Масса полезного груза, кг	800	1819
Масса топлива, кг	9440	9411
Тяга двигателя, кН	267	270
Удельный импульс, с	204	208
Время работы двигателя, с	65	65
Полная длина, мм	14 960	17 955
Диаметр корпуса, мм	1650	1650
Размах стабилизатора, мм	3564	3564
Характеристическая скорость, м/с	1700	1700
Расчетная высота полета, км	100	100

Рис. Михаила Петровского

ТРАЕКТОРИЯ
ПОЛЕТОВ

На схемах цифрами обозначены:
1 — отделяемая головная часть, 2 — отсек
системы управления, 3 — спиртовой бак,
4 — кислородный бак, 5 — двигатель, 6 —
хвостовой отсек, 7 — мортиры с контейне-
рами ГеоФИАНа.

The V-1A and V-1E geophysical rockets. The illustration was created by a team under academician Vasili Mishkin, Hero of Socialist Labor, and Pilot and Cosmonaut of the USSR Vladimir Aksyonov, twice Hero of the Soviet Union. Collective consultant: State museum of the History of Spaceflight K.E. Ziolkovski

Legend:
1. separable nose section
2. guidance system package
3. alcohol tank
4. oxygen tank
5. engine
6. tail section
7. connection to the GeoFIAN container

V-1A and V-1E Geophysical Rockets: General Technical Information		
Geophysical Rockets	V-1A	V-1E
Launch weight, lbs	30,666	31,329
Payload, lbs	1,763	4,010
Fuel weight, lbs	20,811	20,747
Engine thrust, lbs	60,023	60,698
Specific impulse, sec	204	208
Running time of main engine, sec	65	65
Length of rocket, ft	49	59
Maximum diameter of rocket, ft	5.41	5.41
Span over fins, ft	11.7	11.7
Typical speed, ft/sec	5,577	5,577
Calculated flight altitude, mi	62	62

In the summer of 1950, the 22nd Special Purpose Brigade carried out a tactical exercise with R-1 rockets at the Kapustin Yar rocket range.

After all the tests that had been carried out, by order of the Council of Ministers of the USSR on November 25, 1950, the R-1 rocket was added to the arsenal of the Soviet Army with the complex of ground equipment (troop index 8A11). On November 30, 1950, the R-1 rocket was taken on strength by the first rocket unit—the 22nd Special Purpose Brigade, detached to the rocket range at Kapustin Yar. It was given the NATO reporting name SS-1 Scunner.

In December 1950, a second rocket unit, the 23rd Special Purpose Brigade, was formed (Kamyshlov, Stalingrad area). At the beginning of the 1950s, nine special purpose brigades of the high command (22nd, 23rd, 77th, 90th, and others) were equipped with the R-1 rocket.

By Order 380 of the Minister of Equipment of the USSR issued on June 1, 1951, production of the R-1 rocket from domestic components began at Factory 586 in Dnepropetrovsk (later SKB-586 and KB Yuzhnoye).

A summary of rocket production:

1951: 70 examples
1952: 230 examples
1953: 700 examples
1954: production was raised to 2,500 during the year.

Assembly of the rockets from components and parts produced in the factories of Research Institute 88 and OKB 456 was originally carried out at Factory 586, chief designer V.S. Budnyk. The first rockets produced at Factory 586 were sent to the Kapustin Yar rocket range in June 1952. The first successful launch of a rocket from the production line took place at the rocket range on November 28, 1952. Development of the R-1 rocket required the cooperation of thirteen design bureaus and thirty-five factories.

Project development to modernize the R-1, raising the production technology level and improving its effectiveness, began at Dnepropetrovsk in 1953. In the end, the R-1M rocket differed from the R-1 prototype in its simplified design and improved control system, which doubled the rocket's accuracy. In 1955, flight trials of the R-1M were successfully completed after ten launches. The rocket never entered production, however, as it did not satisfy the growing requirements of the military, especially with respect to range.

Unloading an R-1M missile.

Comparison of Tactical-Technical Characteristics of the R-1 First Series and Production Rockets		
	R-1 First Series	R-1 Production
Length	46.75 ft	46.8 ft
Diameter of rocket body	5.4 ft	5.4 ft
Size of stabilizers	11.70 ft	11.70 ft
Launch weight	28,748 lbs	29,608 lbs
Weight of fuel components, Hydrogen peroxide and gases	20,822 lbs	20,743 lbs
Empty weight	8,818 lbs	8,851 lbs
Payload	1,796 lbs	2,370 lbs
Weight of warhead		1,730 lbs
Maximum firing range	167 mi	167 mi
Maximum altitude of flight path	47.85-51 mi	47.85 mi
Maximum velocity at end of powered flight	4,593-4,921 ft/sec	4,806 ft/sec

As per General Staff Directive 2/65616 of June 23, 1954, between July 10, and November 30, 1954, the 233rd Independent Howitzer Artillery Brigade was reorganized into the 233rd Wearer of the Order of Bogdan Khmelnitski Engineer Brigade of the High Command Reserve (T/T 33602), with four battalions as per formation plan 8/494. It was deployed to the city of Klintsy in the Briansk district (Voronezh Military District) with two 8K11 battalions and two BMD-20 battalions (thirty-eight rockets; by December 5, 1954, the BMD-20 battalions handed in all of their equipment, one battalion took part in an exercise near Totskoye in August 1954). The brigade was immediately attached to the deputy of the Commander of Artillery of the Soviet Army. Ground equipment for the launching of 8K11 rockets was issued for the training of personnel for two battalions by December 5, 1954. The brigade received the second part of its equipment after the technical maintenance facilities were ready.

In September 1954, the third battalion of the 233rd Engineer Brigade took part in an experimental exercise at the Totskoye training grounds in which nuclear weapons were used.

As per Directive of the Chief of the General Staff of the Soviet Army 3/464128 of May 1955, the 233rd Engineer Brigade was reorganized into three independent brigades. Unit numbers were assigned and unit banners issued. From July 1960, to April 1963, the third battalion was removed from the rocket force and existed as an independent unit within the 1st Tank Army. The conversion probably took place at the 8K11 complex.

With the arrival of R-1 rockets, the brigades ceased to be special purpose brigades of the high command. In 1958, the 77th and 90th Brigades became part of the land forces.

Conforming with documents from the years 1949 to 1950, the R-1 rockets were accepted for use against large military-industrial targets, important administrative-political centers, and other targets

A missile battery on the move.

of strategic or operational significance. The special purpose brigade could also be used by battalion as part of frontline operations, with rail transport to their deployment area, which would be eighteen to twenty-two miles from the front line. The planned firepower of the brigade with R-1 rockets was supposed to be twenty-four to thirty-six launches per day, while that of an independent battalion was eight to twelve rockets per day.

In 1956, the R-1 missile began leaving the arsenal and was replaced by the R-2 and other missiles. By 1957, 296 test ignitions of R-1 engines and seventy-nine launches were carried out at the Kapustin Yar rocket range. By 1960, the R-1 rocket had left the inventory of the Soviet Army.

Historic Significance of the A-4 and R-1 Rockets

Altogether, sixteen years passed from the start of development work until the existence of comparatively reliable rocket systems. Of these sixteen years, seven took place in Germany (Peenemünde), two years can be judged as joint German-Russian activity (Bleicherode), and seven years were purely Russian activities. In this regard the R-1 rocket with its entire complex of ground equipment has a record for the length of its general development cycle.

One must not underestimate the historic significance of the A-4 and R-1 rockets. They represented the first breakthrough in a completely new technological field: the creation of large, complex technical systems integrating many scientific disciplines and the most varied technologies. Neither the Germans nor the Soviets began with a theory or practical experience. Both in Germany and the Soviet Union, the apparatus and totalitarian leadership of the state created the most favorable conditions for this work to be carried out. As well, in Hitler's

Germany and later in the Soviet Union, all participants in the project were asked to shorten the development cycle as much as possible. And that was no less than sixteen years!

None of the subsequent development programs for much more complicated and complete rocket armament systems exceeded a period of six to eight years. It was not the terrible orders of the state leadership that determined the development cycle, instead it was the experience and knowledge of the academics, the engineers, and all of those who participated in the development of these major systems.

The strategic importance of the R-1 rocket lay not in its military capabilities; instead, its greatest role was serving as a teaching tool for many designers, scientific and testing centers, rocket production organizations, the military and civilian specialists under various authorities, and ultimately for the development of a mighty missile infrastructure in the Soviet Union. More than 200 A-4 and R-1 rockets were launched between 1947 and 1949.

With development of the R-1B, R-1V, R-1D and R-1E geophysical rockets began the separation of scientific space research from the use of rockets as a military system.

The Kapustin Yar missile test range, far from the missile launch sites.

THE R-2 MISSILE

Design work on the R-2 missile (NATO codename SS-2 Sibling) with a range of 360 miles, which later received the designation 8Ж38 (8SH38), began in 1946, while the Soviet specialists were still in Germany. The work was based on the results of experiments on the liquid-fuel engine of the German A-4 rocket. The engine had tremendous power reserves, which made it possible to increase thrust to 70,548 and even 77,160 pounds, with a standard thrust of 55,155 pounds.

There were four variants of the project with the objective of achieving the desired range with a non-separable warhead while retaining the outer dimensions of the A-4. The fifth variant, which was accepted as the basis of the new missile, saw the cylindrical part of the missile lengthened by 6.25 feet. Everything else remained unchanged, apart from the capacity of the tanks.

Despite the extremely tight deadline, the complete set of technical drawings and the explanatory report were prepared by the end of 1946, and three prototypes of the R-2 missile were even completed.

Defense of the R-2 design project took place in April 1947, at a meeting of the Scientific Council of Research Institute 88, with Minister of Armaments D.F. Ustinov in attendance. The decisions that were taken, based on the copies of the German missile, acknowledged the inadequate durability of the missile body. In this context it was decided to fundamentally alter the design scheme of the R-2 missile: to separate the nosecone at the end of the powered phase of flight. This potentially made it possible to dispense with

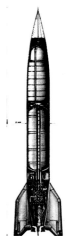

Design of the R-2 missile.

the fuel tank's protective cover and also simplify the tail section—eliminating the stabilizing fins.

The R-2 project was worked out at the end of 1947; however, the new scheme was only partially realized: it was limited to load-bearing propellant tanks, left the protective cover over the oxygen tank and the aft section with fins. An integral oxygen tank was added to the final version. Therefore the initial version of the rocket with an integral propellant tank and simplified control system was separated in a special program. This missile was given the designation R-2E. The E derived from the Russian experimental naya. The following successful trials with the R-2E missile contributed to adjustments in the R-2 missile's operating diagram, and in the ultimate variant it also had an integral propellant tank.

The use of aluminum alloy had an important significance for the new R-2 missile. The partial use of such alloys reduced the rocket's empty weight significantly: the R-2 was a total of 770 pounds heavier than the R-1 although it had twice the range.

The lower arrangement of the sealed equipment module, below the LOX tank, was standard for the R-2 missile. The control system also included a radio command guidance system for improved accuracy.

A new engine—the RD-101 single-chamber, liquid-fueled cruise engine (8Д52)—was developed by OKB-456 led by Valentin Glushko based on the technical solutions of the RD-100 engine. The propellant was made up of ethyl alcohol and liquid oxygen.

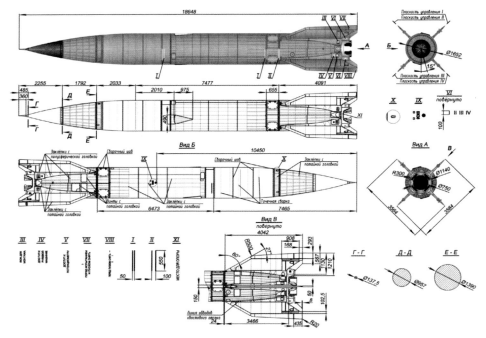

R-2 (8Ж38) long-range ballistic missile.

Its characteristics, however, were significantly improved: length was reduced to 137.8 inches at the cost of changing the arrangement, and most importantly, its weight was reduced to 2,050 pounds compared to the 2,083 pounds of the RD-100 engine while simultaneously raising thrust by 22,050 pounds, specific impulse by four seconds and its capabilities by more than twenty seconds.

Combined with the increase in range, the R-2 missile was equipped with the system for emergency shutdown of the engine. Loss of guidance outside the rocket range could have had unforeseen consequences.

The new Don telemetric system, with twelve continuous and twelve discrete channels, the distance measuring system as the source for the detection devices, and the speed measurement system, were also used for the R-12 missile.

Simultaneously the ground equipment was also perfected. There was such an organic link with the rocket that everyone began calling this system a rocket complex. Use of a new fueling method, in which the oxygen would be displaced by compressed air, was planned for the tests with the R-2E missile. For completion of work on the R-2 rocket in preparation for launch, a vertical test stand was built at the N-88 facility (to simulate launch and flight). It also enabled the complex testing of the links between the missile's on-board guidance system and the automatic engine systems.

RD-101 liquid-fuel rocket engine that powered the R-1 ballistic missile.

Combustion chamber of the RD-101 engine.

Transporting the missile's centers section.

R-2E Rocket: General Technical Information	
Suborbital ballistic short-range booster rocket. The R-2E prototype tested technological innovations for production of the R-2 missile.	
Failures	2
First failure	October 2, 1949
Range	358 mi
Accuracy when fired at maximum range	± 5 miles in altitude ± 2.5 miles in azimuth
Length	55.6 ft
Maximum diameter	5.4 ft
Maximum weight (gross)	44,753 lbs
Weight of fuel components	35,097 lbs
Weight of separated nose section	2,910 lbs
Terminal velocity	7,120 ft/sec
Maximum altitude	105 mi
Fuel components	liquid oxygen and ethyl alcohol
First launch	September 25, 1949
Final launch	October 11, 1949
Launches	3

Missile Complex

The development of the R-2 missile resulted in a new term: missile complex. A complex included various missiles, but also various launch installations such as the mobile launch rail, the shaft, or open launch platform. The development of a new missile also resulted in a new missile complex. The term "rocket complex" includes everything associated with transport (e.g. mobile launch rail), the launch preparation systems, the launch itself, and the missile.

In September 1949, all the work associated with preparing the R-2E missile for flight testing was completed. Improved tests were temporarily carried out in the forest not far from the Podlipki station in July 1949. The test program incorporated virtually all pre-launch operations up to engine start.

Flight testing of the R-2E missile took place at the state central rocket range in September–October 1949, and two of the five launches were failures. In each case the telemetry system on board the missile, with a for-that-time high informational content, made it possible to determine the reason and location of the failure. The reasons were different, but they were not related to the new design solutions incorporated in the R-2. This made it possible to quickly correct the technical documentation and carry out follow-up work on the rocket's equipment. Several small items that had not been completed were identified during the flight tests and were completely natural for the first tests with a new type of rocket and had no effect on the state commission's generally positive assessment of the test results. This made it possible to turn to the state authorities with a proposal for the production of a new series of R-2E missiles (with the necessary follow-up work) and representatives of the Ministry of Defense concerning a test program.

It was envisaged that two batches of missiles (up to fifteen missiles in each) would be produced to achieve the ultimate R-2 configuration, as several new types of guidance system were to be tested in variants of the R-2E to ensure that the level of accuracy promised by the R-2 was achieved. In this way the R-2 missile was to begin a new era, not only in developing the missile's original design circuitry, but also in the area of developing the original circuitry of the guidance system, which ensured qualitative changes needed to improve accuracy.

The two-stage shutdown of the engine was envisaged for this goal, which promised a significant reduction in the scatter caused by follow-up pulses when the engine was shut down.

The R-2 missile differed considerably from the R-2E in the design of individual elements: it was envisaged that the tail sections of some R-2 missiles would be made of aluminum alloy, resulting in a significant weight saving (about 550 pounds). Special attention was paid to the reliability of the missile's communication feed and prevention of an engine fire, which had been observed during testing of the

Comparison of Technical Data of the R-2E Experimental Missile and Production Versions of the R-2 Missile		
	Experimental R-2E	Production R-2
Length	55.6 ft	57.4 ft
Diameter of missile body	5.4 ft	5.4 ft
Dimensions of stabilizing fins	11.7 ft	11.7 ft
Launch weight	44,754 lbs	45,010 lbs
Weight of fuel and other Components	35,097 lbs	34,277 lbs
Warhead weight	2,976 lbs	3,306 lbs
Range	358 mi	366-372 mi
KBO	4,101 ft	
Maximum deviation from range	± 5 mi	
Maximum course deviation	± 2.5 mi	
Velocity at end of powered flight	7,119 ft/sec	
Maximum altitude	105 mi	
Flying time	462 sec	
Weight of explosives	2,222 lbs	
Weight of oxidizer	20,073 lbs	
Weight of fuel	14,204 lbs	
Empty weight	9,982 lbs	
Type of warhead	separable	
Explosive	Trotyl explosive; 8W12 igniter	
Weight of warhead	3,307 lbs	
Weight of explosives	2,222 lbs	
Size of zone of maximum Destruction	10,225 ft²	

Geran: warhead with radioactive liquid, tested from 1953 to 1956
Generator: warhead with radioactive liquid in projectiles, tested from 1953 to 1956.
RDS-4: nuclear warhead, yield 10 kilotons, not used with production missiles, tested in 1956

R-2E missiles. The R-2 was also to be used in special research for development of the electronic guidance system, which was preferred for the next generation R-3. The R-2R was built for this purpose, and the first of this type was launched on October 15, 1952. A total of six R-2R launches were carried out, the last on January 1, 1955.

Flight trials with the first batch of R-2 missiles took place at the State Central Rocket Range in October–December 1950. Twelve were launched and all were failures: five failed during the powered segment of flight due to failure of the guidance system and engine and the unsealing of pipes, and effects of overheating during the descent phase were observed on the nose sections of seven missiles.

Measurements taken during the tests showed that the intensity of vibrations for the missiles with a duralumin tail section was much higher than for missiles with a tail section made of steel. Laboratory investigations revealed that the vibrations noticed on missiles with duralumin tail sections led to increased drift by the gyroscope, which registered

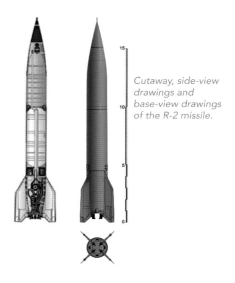

Cutaway, side-view drawings and base-view drawings of the R-2 missile.

vented the destruction of the warhead during the unpowered phase of flight at the cost of increasing its heat shielding.

The measures that were taken (stabilization of the nose section, higher standards for sealing of pneumo and hydro connections, the abandonment of duralumin in the tail section, selection of the variants of the guidance system, and the improved reliability of the onboard wiring system) were to ensure a sufficiently high degree of operating and flight characteristics for the R-2. Of thirteen R-2 missiles launched during the second series of flight tests, twelve reached their targets and just one launch failed because of a production error.

There followed a batch of R-2 missiles earmarked for control tests and to determine if the R-2 could be added to the Soviet Army's arsenal. Sixteen missiles were ordered, two of which were used for

the missile's actual position on the flight path. The result of this drift led to a development that destroyed the missile.

In the end the duralumin tail section had to be abandoned, and a steel tail section was adopted in its place.

During testing of the first batch of R-2 missiles experiments were carried out that showed that guidance with respect to the longitudinal axis in the vertical plane was possible without aerodynamic controls. This made it possible to simplify the guidance circuitry in the second batch of R-2s. Technical solutions were found that pre-

training the personnel of the special purpose brigades. The tests were carried out at the State Central Rocket Range in August–September 1952. Of fourteen missiles, twelve completed their missions. As a result of these tests, in 1952, the R-2 missile and its complex of ground equipment were added to the arsenal of the Soviet Army with the troop designation 8SH38. That year production began in factories N-88 and N-586.

Simultaneous to the redesigning of the R-2 missile and preparations for its adoption by the army, work was carried out to standardize the equipment of the ground complexes for the R-1 and

Frequency (vibration) tests on the R-2 missile at the ZNIIMasch (Central Scientific Research Institute of Mechanical Engineering), 1951.

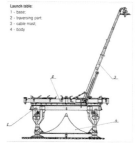

Launch platform for the R-2 missile. Launch cable, 1. base, 2. transverse section, 3. cable mast, 4. body.

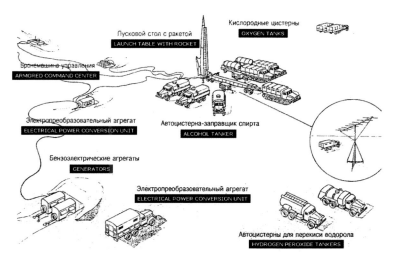

Пусковой стол с ракетой
LAUNCH TABLE WITH ROCKET

Кислородные цистерны
OXYGEN TANKS

Бронемашина управления
ARMORED COMMAND CENTER

Электропреобразовательный агрегат
ELECTRICAL POWER CONVERSION UNIT

Автоцистерна-заправщик спирта
ALCOHOL TANKER

Бензоэлектрические агрегаты
GENERATORS

Электропреобразовательный агрегат
ELECTRICAL POWER CONVERSION UNIT

Автоцистерны для перекиси водорода
HYDROGEN PEROXIDE TANKERS

Frequency (vibration) tests on the R-2 missile at
the ZNIIMasch (Central Scientific Research Institute
of Mechanical Engineering), 1951.

R-2 missiles. The development of new equipment for the R-2 missile's ground organization improved its employment and operational characteristics. To shorten time spent working on the missiles in the technical and launch positions, two special machines were introduced. One, with the KUNG superstructure, carried spare parts, tools, and equipment and formed part of the ground complex's equipment. The EWA machine was subsequently used at other missile complexes, significantly improving their usage characteristics. In May–June 1954, control flight tests were carried out with the R-2 to check over and modify the entire complex of technical documentation envisioned for mass production. Ten R-2 missiles were tested, of which eight completed their missions.

Investigations were also carried out into firing the missiles over shorter ranges (120 and 160 miles). During launches over shorter ranges, greater deviations from the target were observed, caused by the delayed separation of the warhead. A modified version of the R-2 with a heavier nose section was used for subsequent shorter-range tests. Flight tests with this version in July–August 1955, were successful.

From 1953 to 1956, R-2 missiles were also launched from the State Central Test Range carrying special warheads (codenamed Geran and Generator). The Geran warhead was filled with a radioactive liquid. Exploded at great height, the liquid was supposed to fall as a deadly rain. Generator differed from Geran in that the same radioactive liquid was carried in many small containers, with each of these containers exploding independently above the earth.

During preparations to launch the first Geran, a cloudy liquid ran from the warhead of the rocket, which was sitting on the launch platform. The tank with the deadly liquid had sprung a leak. The entire launch crew was able to hurriedly leave the launch site. And what was to become of the rocket? Voskresensky, who never lost his presence of mind during a launch, slowly approached the rocket. Watched by the crew that had fled and was now about 300 feet away, he climbed onto the rocket's tail section, so that everyone saw it, theatrically raised his hand, and ran his finger upward through the liquid running from the rocket body. Then he turned to face the stunned onlookers, stuck out his tongue, and licked the "radioactive" finger. Then Voskresensky came down, moved away slowly, and said: "Men, go back to work. It's harmless dirty water."

He was convinced that this liquid only simulated the atomization process and he was not wrong. Nevertheless, that evening in the hotel he drank an additional portion of alcohol "to neutralize the worry he had experienced."

The flight control tests involving the first domestic ballistic missiles, the R-1 (8A11) and R-2 (8SH38), had the following main results:

Completion of the redesigning of the missile's onboard systems and also its ground equipment.

The method of preparing the missiles for launch was created and tested in practice; new technical schematics of the launch preparations were developed.

The interaction between all organizations taking part in the flight test program was revised. Cooperation between all services at the test range was considerably improved.

Empirical data was gathered about the working of the missile's onboard systems during flight; actual data was gathered on the range of the warhead and its deviation from the target. The methodology for making ballistic computations was refined.

The contents of the operating and technical documentation, necessary for checking out the missile's equipment and its preparation for launch, were refined. New documents were developed for the organization (technological plans, budgetary test orders, calculation document forms, missile logs, etc.), which provided a basis for training personnel. Also modified was the plan for introducing design changes into production and directives for flight testing from the builder (directions for follow-up work, confirming its completion, additional fixes, etc.).

New military organizations, the special purpose brigades, were created in the Defense Ministry of the USSR.

All of the experience gained during testing of the domestic R-1 (8A11) and R-2 (8SH38) missiles was used by other organizations interested in other applications for missile technology.

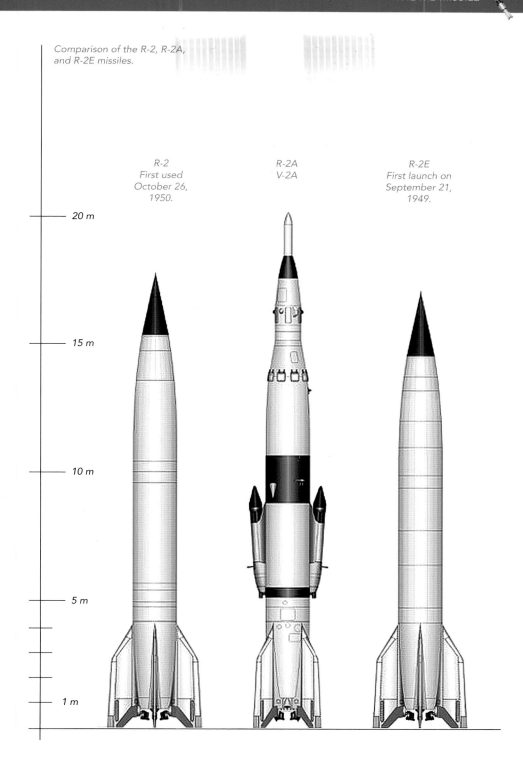

Comparison of the R-2, R-2A, and R-2E missiles.

R-2
First used
October 26,
1950.

R-2A
V-2A

R-2E
First launch on
September 21,
1949.

20 m

15 m

10 m

5 m

1 m

Modifications

R-2E: Experimental missile with a simplified guidance system (tested September 1949). The second variant of the design project, on which work began in 1947.

R-2/8SH38: R-2 base version and production service missile. The first series of missiles was tested from October to December 1950.

R-2R: Research vehicle with a system of radio-controlled correction of the flight path. It is probable that this radio guidance system was later updated for the R-3 and later the R-5 missiles. Development of the radio guidance system began in June 1950. Testing of the missile with the radio guidance system continued until May 1951.

R-2 with the RD-103 Engine: Burn tests with the R-2 missile with the RD-103 engine were carried out on the test stand in Sagorsk in 1952. The engine was later adapted for the R-5 missile.

R-2/8SH39: Modification of the R-2 missile (the designation was mentioned in a number of mass media) with heavy warhead. The missile was designed for improved accuracy against closer-range targets (120 to 180 miles). Successful trials in July-August 1955.

R-2M: Test missile with РДС-4 nuclear warhead. Tested in 1956. No series production.

R-2A (W-2A)/R-2B (W-2B): Geophysical missiles; launches began in 1957 (R-2A – May 1957 – two launches; August-September 1957 – three launches; August 1958 – two launches)

R-2A Missile Specification:

Weight of nose section: 3,086 pounds

Weight of container with scientific apparatus: 573 pounds

Height of ascent: 130 miles

RD-103/RD-103M Engine Specification		
	RD-103/RD-103M	M8Д54/8Д71
Development Period	1952–1953/1955	
Propellant components – Oxidizer – Fuel	liquid oxygen 92% solution of ethyl alcohol	
Thrust – on ground, lbs – in a vacuum, lbs	94,798/97,003 110,231/112,436	
Specific Impulse – on ground, sec – in a vacuum, sec	220 248	
Pressure in combustion chamber	342-346 psi	
Weight of engine: – dry weight, lbs – in operating condition, lbs	1,918/1,911 lbs 2,270 lbs/-	
Dimensions – height, ft – diameter, ft	10.24/10.24 ft 5.41/5.41 ft	
Operating time, sec	115/120	

WR-190: Project for a manned geophysical missile (end of the 1940s).

The planned organization of the special purpose brigades equipped with the 8SH38 missile called for three firing battalions with two launch batteries in each (total of six launch sites in the brigade). Manning of the complexes: eleven persons.

Status: Entered service with the Soviet Army beginning in 1956. In 1960, the missile complexes were taken from the army. There was no export.

The second type of domestic ballistic missile was accepted into service in 1951. It was designed by NII-88's SKB-1 (special design bureau) under chief designer S.P. Korolev. Compared to the R-1 (8A11) missile, it had a radio guidance system that improved accuracy, a warhead that separated after engine burnout, and more than twice the range.

The R-2A on the launch platform.

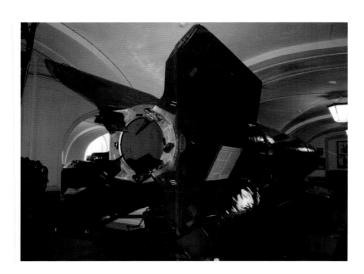

The R-2 missile, Museum of Artillery, Engineer and Communications Troops, St. Petersburg.

Comparison of Technical Information for the R-1 (8A11) and R-2 (8SH38) Missiles		
	R-1 (8A11)	R-2 (8SH38)
Year added to the arsenal	1950	1951
Range, mi	31-167	124-278
Rocket length, ft	47.24	58.4
Maximum diameter, ft	5.4	5.4
Span over stabilizing fins, ft	11.7	11.7
Weight of fueled missile, lbs	29,541	45,018
Weight of propellant (spirit), lbs	9,039	
Weight of oxidizer (liquid oxygen), lbs	11,177	
Weight of warhead, lbs	2,394	3,349
Weight of explosives, lbs	1,720	2,222

Export to the People's Republic of China

End of the 1950s. Design documentation for the production of R-2 missiles and probably sufficient components for several missiles were passed to the People's Republic of China. The missile was introduced into service by the People's Liberation Army under the designation DF-1.

The first R-2 missile, delivered from the Soviet Union, was launched from the Jiuquan test site on September 1, 1960.

The Dongfeng 1 / DF-1 / East Wind 1 flew on November 5, 1960. A copy of the R-2 missile (the Model 1059) was built in the People's Republic of China.

Above: The Dong Fend 1 missile on the launch platform. From the mpleio Archive

Right: The Dong Feng 1, a Chinese-built version of the R-2 missile. From the mpleio Archive

THE G-2, G-3, G-4, AND G-5 MISSILES

In September 1946, Gröttrup designed a two-stage long-range missile with a range of 1,500 miles. The idea was picked up again in 1949, and under Gröttrup's leadership the G-2 (R-12) missile was developed, possessing a range of 1,500 miles and carrying a warhead of at least 2,200 pounds. It was intended that this missile should follow directly after the R-10. It was planned to cluster three R-10 engines and thus increase total thrust to more than 220,000 pounds. Jet rudders were dispensed with to avoid their gas dynamic drag. Thrust changes would be used to control the three engines.

Many alternative configurations (R-12A to R-12K) were considered by the German team in the Soviet Union.

These included cluster and sequential engine arrangements, gimbal-mounted engines, and other innovations. The R-12K was particularly interesting, because this concept was later used on the American Atlas missile—jettisoning of the two outboard engines at altitude to increase range.

The G-2 missile was given the secret designation R-6 and the overt designation R-12.

More than ten G-2/R-12 variants were determined in detail, seeking the optimal technical solution. All used a common wall between the aft tank containing fuel and the forward tank with liquid oxygen. Two important aerodynamic shapes were considered—the conical shape, which was aerodynamically stable and independent of speed,

Various technical solutions, from left: a G-1 cluster; R-12A dual sequential phase; R-12 b single-stage; R-12C dual stage; R-12E dual stage; R-12G and R-12H single-stage; and R-12K with jettisonable booster engines.

and a missile with its center of gravity in the middle and cylindrical stages with fins.

The following variants were examined:

G-1 Cluster: Two G-1 missiles as the first stage, flanking a single G-1 as the second stage. Launch weight 110,000 pounds, thrust at liftoff 220,000 pounds, total circumference 19.7 feet, length forty-nine feet.

R-12A Dual Sequential Phase: Conical shape, with three G-1 engines in the first stage and one G-1 engine in the second stage. Length eighty-three feet, diameter at the base thirteen feet.

R-12B Single Stage: Short-range missile to replace the V-2, based on the second stage of the R-12A, diameter 7.9 feet, length 52.5 feet.

R-12C Dual Phase: A variant of the R-12A with a larger diameter upper stage. Sixty-three feet long.

R-12E Dual Phase: Cylindrical first stage with aerodynamic stabilizing fins.

R-12G: Single-stage design with conical shape.

R-12H: Single-stage with cylindrical shape and aerodynamic stabilizing fins.

R-12K: With jettisonable boosters.

The optimal technical solution appeared to be the R-12G missile with its single-stage conical design. In a comparison of the G-2 with Korolev's R-2, the state commission preferred Gröttrup's design, but the Russian designer convinced the government that the innovations of the G-2 exceeded the limits of Soviet technology. The G-2 project got no further than calculations on paper. Korolev's R-2, a modest upgrade of the V-2, was selected for production instead.

On April 9, 1949, Ustinov issued a directive to the German missile engineers to develop a "3000/3000" missile. The objective was to transport a 3,000-kilogram (6,600 pound) nuclear warhead a distance of 3,000 kilometers (1,860 miles), which would be sufficient to reach the United States.

Dr. Albring, the German aerodynamicist, designed the G-3 missile in October 1949. The G-1 designed by Gröttrup was to form the first stage. The second stage was to have a layout similar to that of the missile-powered suborbital bomber designed during the Second World War by Saenger and Bredt. Reaching an altitude of eight miles, the supersonic missile would carry a 6,600-pound warhead a distance of 1,800 miles.

This was an alternative approach to Ustinov's request of April 1949, for the delivery of a 3,000-kilogram (6,600 pound) nuclear warhead over a distance of 3,000 kilometers (1,860 miles). This design was to be reworked in Korolev's design bureau, becoming the experimental winged missile (EKR) with ramjet propulsion (1951–1953). Alternative designations chosen for the G-3 missile were R-8, the secret designation, and R-13, the open one. The designations G-5 and E-15 were also mentioned, but they could have been related to another design.

G-2 Missile (R-6; R-12) Specification	
Maximum range	1,553 mi
Status	1948, project
Launch weight	110,231 lbs
Payload	2,200 lbs
Standard warhead	2,200 lbs
Length	80.4 ft
Diameter	12 ft
Thrust	224,808 lbs

G-3 Missile (R-8; R-13) Specification	
Maximum range	1.553 mi
Status	1949, project
Launch weight	88,184 lbs
Empty weight	6,085 lbs
Payload	7,495 lbs
Standard warhead	7,495 lbs
Length	91.85 ft
Diameter	5.4 ft
Thrust	72,838 lbs

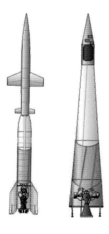

Cutaway drawing of the G-3 (left) and G-4 missiles.

The ED-140 engine.

The G-4 missile was designed by the German team under Gröttrup in the Soviet Union to compete with Korolev's R-3. A multitude of technical solutions were considered, including a cluster (or packet as the Russians would later call it) of three G-1s; missiles with wings analogous to the development of the A9/A10 or the Saenger suborbital bomber (the G-3); and sequential stages, as used in the earlier G-1 and G-2 designs. Also considered were balloon tanks of high-grade steel (as used in the USA on the Atlas ICBM). But the Russians did not have such alloys, and the Germans kept the idea to themselves "in reserve."

The Gröttrup team completed the twenty-volume design study in June 1949, three months after it was given the green light. The selected configuration was a single-stage cylindrical design, which was aerodynamically stable in all flight conditions. Like the earlier designs, the liquid oxygen tank was in the forward position compared to the V-2, overcoming pre-cooling problems for equipment and engine. The warhead was in a cylinder with a diameter of 4.6 feet. The reentry section used the "fast point" pointed nose preferred by the Soviet engineers. For their part, the Germans preferred the "slow point" design—a blunt nose with conical sides. But the Russians feared that a slow-falling warhead could be intercepted and believed that the thermal problems of the pointed nose could be solved.

A new cylindrical high-pressure combustion chamber was used for the R-14. It fed a spherical mixture chamber. The German engineers worked with Glushko to build an experimental subscale chamber that produced 15,432 pounds of thrust, burning liquid oxygen and kerosene at a pressure of sixty atmospheres. Nineteen of these chambers were used as combustion chambers in Glushko's RD-110 engine, which was planned for use in Korolev's competing R-3 missile. According to Russian sources the ED-140 engine was tested one hundred times between summer 1949, and April 1950.

Unknown to the German team, their approach was evaluated against Korolev's R-3 on December 7, 1949—and their design was found to be superior. Neither ended up in production, but the design concepts of the G-4 led directly to Korolev's R-7 ICBM (essentially a cluster of G-4 or R-3A elements). Notable unique features of this missile derived from the R-4 were its conical shape, the "sharp point" shape of the warhead or payload shroud, forward location of the oxygen tank, and elimination of aerodynamic surfaces.

Work on the G-4, the last design by the German team, continued until 1952. The G-4 was designated as the R-10 in the original secret Russian numbering scheme and was known to the Russians as the R-14. The same designation was given to the 1,800-mile-range single stage IRBM finally delivered by Yangel to the Soviet armed forces in 1962.

Another project for a carrier of a nuclear payload, which appeared under the designation "G-5" ("R-15"), turned out to be a modified variant of the R-4 ballistic missile. Based on the characteristics contained in the technical specification, it was comparable to Korolev's prospective "R-7" missile.

Several sources suggest that the G-5/R-15 intercontinental ballistic missile was designed by the Gröttrup team. If it was, then the "package of G-4 missiles" was the direct forefather of Korolev's R-7.

The designation G-5/R-15 was also used for a missile with ramjet propulsion as the G-3 or R-13.

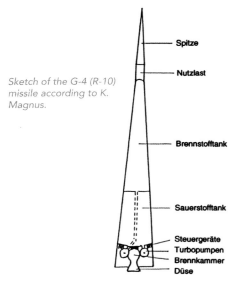

Sketch of the G-4 (R-10) missile according to K. Magnus.

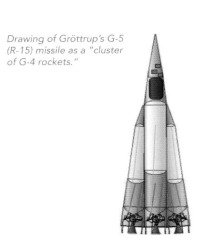

Drawing of Gröttrup's G-5 (R-15) missile as a "cluster of G-4 rockets."

G-4 (R-10; R-14) Missile Specification	
Maximum range	1.864 mi
Status	1949, project
Launch weight	55,115 lbs
Payload	6,600 lbs
Standard warhead	2,200 lbs
Length	82 ft
Diameter	12.24 ft
Thrust on ground	222,560 lbs
Thrust in a vacuum	238,972 lbs

G-5 (R-15) Missile Specification	
Maximum range	1,864 mi
Status	1949-1952, project
Payload	6,600 lbs

GORODOMLYA ONCE AGAIN

As early as 1949, the German missile experts began losing faith in their work. They worked on the projects given them without having the opportunity to exchange information with the Soviet missile specialists. Their work was top secret, so there was no possibility of the German and Soviet experts sharing information on common themes.

Working to a plan was quite a new experience for the Germans. Dr. Albring formulated it thus: "We nevertheless had to accept that this for us new method of monthly analysis helped develop working discipline." Under this system a project developed in strict stages, like a stepladder. This method of work was not received with unconditional love by the scientists.

A shallow water canal and later a supersonic wind tunnel were built with modest resources for flow experiments. In mid-1951, considerations were played out on paper, with extensive theoretical breakthroughs for the R-14 program:

- missiles with special propulsion systems,
- the stage principle, and
- clustering of identical engines.

Dr. Umpfenbach tested various combinations of fuels in the combustion chamber. The Germans also worked on development of a system that would enable a vehicle to maintain a certain altitude based on barometric pressure. The device would possibly be required for a type of cruise missile.

But the desire to return home grew stronger. Even the formation of two theater groups, sports festivals, and other cultural events could not hide this fact. At the same time, the assignments given the Germans became increasingly less significant. Resent-ment among the Germans rose, and written requests for their return to Germany became ever louder.

In March 1951, the number of Russian experts on the island was increased. At the same time, the leadership on the island announced to the Germans that twenty of them would be leaving the island by June 3, 1951. The group that was leaving consisted of experts who were no longer needed. Vasilyev, the island's chief designer, announced that: "As of October 1, 1951, only 'peaceful work' will be continued. The military programs are being stopped."

It was not learned until decades later that the Soviet government finally decided in September 1951, to return the German missile experts to Germany. On January 20, 1951, a group of fifty persons, including women and children, left the island. More experts left the island on June 10 and 13. Fifteen percent of the personnel stayed behind—totalling seventy people, with women and children. Repairs were now on the program. There was the opportunity to systematically pick up knowledge, but no answer to the pressing question: when are we going home? The last "German raketchiks" left the island of Gorodomlya for the German Democratic Republic on November 16 and 20, 1953.

The Gröttrup team, still in Moscow, followed on November 28, 1952. The subsequent fates of the German missile experts were very different. Some left the German Democratic Republic, seeing their future in West Germany, like Dr. Magnus and the Gröttrup Family, for example. Others found a home in the GDR, like Dr. Albring, who became a lecturer at the Dresden Technical University. The greatest rise was made by Erich Apel, engineer at Peenemünde, head of workshops on the island. In 1957, he became Minister of Heavy Engineering, member of the central committee of the Socialist Unity Party of Germany. His rapid rise led via the Secretary of the Economy and head of the State Planning Committee to Deputy Premier of the GDR. Dr. Apel took his own life on December 3, 1965, out of desperation over the hopelessness for a new economic policy in East Germany.

A question that is frequently asked is: what role did the German workers play in the development of Soviet missile and space technology?

B.E. Chertok wrote: "The most important accomplishments by the German missile experts were not the work they did during their time in the Soviet Union, rather the work that they did in Peenemünde until 1945. The building of such a tremendous scientific research base as the one in Peenemünde, the development of the A-4 missile system, the mass production of missiles, the beginning of work on perspective long-range missiles, ballistic winged missiles, multi-stage missiles, the development of various types of surface-to-air missiles, especially those like the Wasserfall, that is the foundation, the starting place from which virtually all subsequent work by both the Soviet and American missile builders emanated."

At the Missile Reunion at the Dresden Technical University in 2006, Dr. Albring formulated the accomplishments of the German missile experts of the Gorodomlya group as follows: "After studying the images of the Vostok missile I observed that the Russians had used two design principles, the clustering of 55,000-pound-thrust engines into units of four, and the simultaneous ignition of engines in the first and second stages on the ground during launch."

One reason why the German missile experts did not play an active role in the Soviet space program was the traditional Russian distrust of foreigners, enhanced and multiplied by anti-Soviet western policies.

In January 1949, the Soviet Academy of Science began a campaign against "kowtowing to the west" and confirming the primacy of Russian science. In this sort of atmosphere, naming a German for the top position in a secret Soviet defense project was impossible.

For his part, Ustinov wanted to use the Germans as tutors and practical alternatives to Korolev, in order to stimulate competition.

The technical experience of the German experts of course saved the Soviets many years of exhaustive

Dr. Albring at the 2006 Rocket Meet at the Technical University in Dresden. From the private archive of E. Tremel

work. The Soviet specialists in Germany grasped very quickly that missile technology was not a matter of organization or a ministry, rather a task for the entire state.

The fact that, after the difficult war, the Soviet Union adopted the German achievements and then outdid them in a very short time was of great importance to the general technical culture in the country. The development of missile technology was a powerful stimulus for the development of new scientific disciplines, for example electronics, computer technology, cybernetics, gas dynamics, mathematic modeling, and the development of new materials.

The work of the Soviet specialists created a close intellectual core of specialists from widely different fields of knowledge. This unit, created in Germany, survived even after the return to the Soviet Union, although the people were assigned to various ministries. And this unit existed, not in words and deeds, but despite the sometimes difficult personal relationships between the chief designers, their deputies, the ministers, the military, and the government officials.

And how did the West, which was naturally informed about the activities of the German missile experts, react?

Although the German missile specialists were released to the German Democratic Republic, many, including Helmut Gröttrup, moved to West Germany without much difficulty. Some even moved to the United States, and there continued their work in the field of spaceflight. It was no surprise that the US Air Force's Air Technical Intelligence Center, in cooperation with the British Air Ministry, obviously questioned every German returnee from the USSR.

In August 1952, immediately after the twenty leading scientists returned from Gorodomlya, the Air Technical Intelligence Center organized a special

meeting at Wright Field (now Wright-Patterson Air Force Base) in Dayton, Ohio. Attendees were high-ranking military officers from the USA, Great Britain, and Canada. Based on the reports by the German experts, the committee concluded that the Soviet missile program was comparable with that of the USA and that they were pursuing the development of intercontinental ballistic missiles, similar to the American Navajo, Snark, and Atlas programs.

The Germans stated that while they were in the USSR, they had considered a powerful engine capable of producing up to 265,000 pounds of thrust, that it "was possible but not probable" that the USSR could build missiles that could be powered by two or four of these engines. The gathering also talked about whether a winged missile with two such engines could fly up to 4,400 miles, but that would not be sufficient to attack the USA itself. It seemed unlikely that such missiles could be built in the foreseeable future. Nevertheless, the experts agreed that by 1956, the Soviet Union would be able to deliver a warhead of about 2,200 pounds to the northwestern USA, launched by a two-stage ballistic missile. Finally, they estimated that by 1958, the USSR would achieve the capability of hitting targets anywhere in the USA, with warheads weighing approximately 4,400 pounds. That would only be possible if the Soviet government gave such a program a high priority. The representative of the Vultee Aircraft Corporation (future developer of the Atlas ICBM) called such estimates "very optimistic."

Dr. Ernest G. Schweibert, chief historian of the US Air Force Systems Command, summarized the general conclusions of the meeting in August 1952:

There seems to be little cause for unnecessary concern about a war in which ballistic missiles could be used. To the conference attendees the "missile age" appeared to be quite far off.

THE R-3 MISSILE

Development of the R-3 missile with a range of 1,860 miles was one of the main themes discussed at the government level in April 1947. Various configurations were examined in the design project: single stage, composite, and winged. The single-stage scheme was adopted with an initial weight of 66–70 tons and engine thrust of 264,000 to 308,000 pounds. The design of the missile included a separable nose section, integral load-bearing tanks for the oxidizer and fuel, and removal of stabilizing fins.

Work on the R-3 project began under Korolev's leadership at the end of 1947. Basic research into development of a missile with a range of more than 1,800 miles was carried out. Four basic design variants were considered:

BN: *ballistically conventional, single stage*
A conventional single-stage missile for a range of 1,800 miles and two-stage or with jettisonable tanks for increased range up to 4,800 miles. This was the best solution for achieving a range of 1,800 miles. Increasing range to 4,800 miles would require extensive redesign.

BS: *ballistically composite, multi-stage*
A cluster scheme, parallel arrangement of similar or identical missile stages. This was considered the best alternative for an intercontinental missile. M.K. Tikhronravov had in fact proposed an intercontinental missile, consisting of three R-3s—two outer R-3s as the first stage around an R-3 as the second stage and core of the missile.

KN: *winged conventional*
A missile with wings for increased range, like the German A-4b. Compared to a ballistic missile, however, this required the solution of new technical problems concerning guidance, high-temperature materials, and hypersonic aerodynamics. A winged separable warhead: this warhead would have weighed less than the winged missile but it required extensive research in the same technical areas.

KS: *winged multistage*
A two-stage high-speed cruise missile. Both stages were powered by missile and ramjet engines. One would use the R-3A as booster, in order to reach Mach 3 and ramjet ignition.

Korolev's preferred configuration was the BN with the conventional single-stage design. This was selected in June 1949, within the institute leadership. The design required technical advances in all areas compared to the V-2:

Reduction of the empty weight component by a factor of three compared to the R-2. This was achieved through the use of integral propellant tanks; removal of the aerodynamic stabilizers (orientation of the statically unstable missile with a single gimbal-mounted engine) and a structure made of a new light-aluminum alloy.

Improvement of the engine's specific impulse by twenty-two percent. The engine would be powered by LOX and kerosene, producing a thrust of 265,000 pounds and a specific impulse of 288 seconds. Both Glushko's (RD-110) and Polyarny's (D-2) engine designs were considered. Glushko's engine was an analogy to the V-2 engine – nineteen of the 15,400-pound-thrust combustion chambers, developed by German engineers, supplied a single large "mixing chamber." Polyarny had proposed an engine with unspecified "new fundamentals." Korolev supported the use of Glushko's rather conservative engine.

A separable warhead was also used for improved accuracy: the "high-speed tip" preferred by Korolev was part of the warhead. Testing of this configuration began with the R-1A missile series on May 4, 1948.

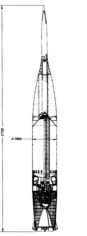

Design project (left) and drawing of the R-3 missile.

Launch weight was seventy-nine tons with an empty weight of 18,700 pounds. The warhead would separate at a velocity of 15,420 feet per second in order to achieve the desired 1,800-mile range. An alternative heavier 13.2-ton warhead could be delivered in a 600-mile area. Barmin's OKB Spez'masch would develop a mobile launcher like that used for the V-2 missile.

The R-3 design project consisted of three main parts:

The first part contained the project—the actual R-3 missile (S.P. Korolev's Department 3 would take over its design. Korolev was simultaneously placed in charge of the other departments of the project).

The second part included the design project and the building of a model of the engine system (its design was decided after a competition between Department 3 of OKB-456 under chief designer V.P. Glushko and the Scientific Research Institute 1MAP under A.I. Polyarny).

The third part comprised all of the problems associated with designing the guidance and control systems (chief designers N.A. Pilyugin and M.S. Ryazanskomu were tasked with their solution).

During preparation of the R-3 missile project, the experience that had been gained in producing the R-1, R-1A, R-2E, and R-2 missiles was summarized and the theoretical and experimental data analyzed. The project developed a sort of stimulus for the creation of the general methodical basis for the development of long-range missiles. Thus the first volume of the project, for which S.P. Korolev was responsible, was given the title Principles and Methods of the Development of a Long-Range Missile. The entire project consisted of twenty volumes and ten additional volumes with calculations by subcontractors. Work on it was completed in June 1949, with assistance from the Gröttrup group. The majority of the project volumes contained two topics: first a generalization for future long-range missiles, and then the second, dedicated to the development of the R-3 missile.

As so many new technologies were involved, it was considered necessary for the R-3A experimental missile to be based on the R-2. It was to serve as a model for the reworking of the exterior conditions of the missile design, the engine, and guidance system of the R-3 missile, the presence and the practicality of experiments in the use of high-energy fuels and energy-rich propellants. The perspective of designing the missile according to the "cluster" concept and the scheme with a separable winged warhead in particular were emphasized. Development of the R-3A experimental missile also benefited the development of perspective design concepts for the research themes planned for the next two years.

The R-3A had a range of 540 to 600 miles with a payload of 3,373 pounds, an empty weight of 9,040 pounds, 45,195 pounds of fuel, and a launch thrust of forty-four tons. The R-3A could also serve as the prototype for a smaller medium-range missile.

The design project was completed by Korolev in June 1949. In November, the project plan was delivered to the NTS (Scientific-Technical Council) of NII-88 for examination. The NTS met for a plenary session on December 7, 1949, and the proposal came under harsh criticism.

Isayev found extensive problems in Glushko's engine development. The enormous increase in thrust, the performance, and use of new fuels seemed to go a step too far. But Korolev insisted that Polyarny's design with older technology would not meet the requirements. In general the council favored Gröttrup's G-4/R-14 design to the same specification. A comparison shows the different performance figures.

After a heated discussion the council approved further development of technology for the R-3 but not the missile itself. This work was to proceed in several directions:

• Building of an R-3A technology demonstrator to be flown as Project N-1. A new medium-range missile with a new lighter atomic warhead could be developed from it (the R-5).

• The RD-110 and D-2 engines would carry out a development test as Project N-2, to prove the LOX-kerosene fuel technology.

• The "cluster" missile and structural research for its use in a medium-range missile was to be continued as Project N-3/T-1.

• Intercontinental winged cruise missile studies were to be continued under Project N-3/T-2.

• Systems development for the long-range systems would be carried out by the Pilyugin and Vavarinka OKB, with parallel work by Konoplev at NII-20. The system used gyroscopic guidance, supplemented by radio-controlled engine shut-down to improve accuracy.

At this time the immediately military requirement to threaten all of Europe with atomic weapons would have been achieved by:

• The development of the R-5 MRBM (3,140-pound warhead with a range of 745 miles) with improved V-2 and R-3A technology.

• The development of the R-11FM submarine-based ballistic missile (2,140-pound payload over a range of 93 miles).

Performance Comparison of the R-3 and G-4/R-14 Missiles		
Parameter	Korolev R-3	Gröttrup G-4/R-14
Payload to 1,800 miles	6,613 lbs	7,495 lbs
Launch weight	156,528 lbs	154,323 lbs
Empty weight	18,695 lbs	15,652 lbs
Burnout speed	15,420 ft/sec	14,764 ft/sec
Specific impulse – vacuum	288 sec	249 sec
Specific impulse – ground	240 sec	234 sec

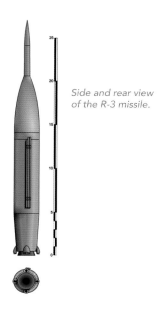

Side and rear view
of the R-3 missile.

R-3 Missile (Project) Specification	
Maximum range	1,864 mi
Status	1949, design
Launch weight	152,119 lbs
Empty weight	12,081 lbs
Payload	11,023 lbs
Standard warhead	11,023 lbs
Length	92 ft
Diameter	9.2 ft
Specific impulse in a vacuum	285 sec
Specific impulse at sea level	244 sec
Thrust at ground level	372,958 lbs
Thrust in a vacuum	307,988 lbs
Burn duration	150 sec

Work on the R-3 project proceeded with a significant German contribution, although the German experts did not know how the work was going. Detailed technical questions were discussed and solved during the entire 1950 – March 1951 period. In the interim Glushko was able to solve the mixture chamber instability in his nineteen-chamber RD-110 engine. Further development of the engines by Glushko and Polyarny was cancelled in 1951.

Work on the R-3A went ahead. The Germans underwent another period of intensive questioning in June 1952. It seemed to them that a test flight by the R-3A was imminent, but in the end it was decided to go with the development of the R-7 intercontinental missile. In the course of development the R-3A missile had reached a range of 580 miles with a payload of just 1,100 pounds. The R-3 project was cancelled in October 1951, however, without ever having flown. The technology used in the project was used to promote development of the R-5 and R-11. Development of the R-3 was totally abandoned.

The single-chamber liquid-fuel engines were developed on the basis of a competition between chief designer V.P. Glushko's OKB-456 (RD-105) and

Comparison of the RD-105 and RD-110/8D55 Engines		
	RD-105	RD-119/8D55
Designer	OKB-456	NII-1 of the MAP
Thrust at ground level	121,194 lbs	308,887 lbs
Fuel	kerosene	kerosene
Oxidizer	liquid oxygen	liquid oxygen
Specific impulse at ground level	260 sec	243.6 sec
Specific impulse in a vacuum	302 sec	285 sec
Note: The engine was developed. Some of the apparatuses were tested. Development was halted because of many technical problems.		

the MAP's (Ministry of the Aviation Industry) OKB-1 under the leadership of A.I. Polyarny (probably the RD-110). Experience gained in developing the RD-101 engine was used by both designers.

The modified RD-101 engine with a thrust of forty-four tons and a specific impulse on the ground of 210 seconds was proposed for the experimental R-3A missile.

Development of the R-3 missile did not proceed beyond the design project stage, but the results achieved at that level exerted an important influence on the subsequent path of spaceflight in the USSR.

The Soviets finally placed a medium-range missile with a range of 1,800 miles into service in 1962, Yangel's R-14.

General Specifications of the R-3 Design Project and R-3 Project Missiles		
	R-3 Design Project	R-3 Project
Length	88.9 ft	53 ft
Diameter of missile body	9.2 ft	6.6 ft
Weight	143,300-156,528 lbs	51,588 lbs
Empty weight		approx. 8,819 lbs
Fuel weight	132,277 lbs	
Warhead weight	up to 6,613 lbs	
Range	1,864 mi	580 mi

NUCLEAR WEAPONS AND MISSILES

In the field of weapons development, the Second World War ended with essentially new types of weapons technology: atomic weapons, radar, and guided weapons. The leading military officers of the Soviet Union became aware of this, as did Stalin's politburo. New priorities had to be created in the defense industry.

Wartime experience had shown that changes were even necessary to conventional weapons systems: target acquisition by radar for the artillery, detection of aircraft by radar, aircraft with the ability to carry and drop atomic bombs, and guided missiles. Modernization of the armaments industry was necessary. But from where was the Soviet Union to get the means and the money? Savings would be found in the branches of industry producing basic human requirements: light industry, the food and automobile industries, and the production of agricultural machinery.

It was obvious that in addition to the economy and science, government organization required revamping. A special government agency was required, led by a member of the politburo, who would be directly accountable to Stalin but would also be able to make non-bureaucratic decisions necessary for development of the new technology.

The first government agency was the Soviet Radar Council. Even during the war this problem was clear. On July 4, 1943, the "decision to found the radar council of the state defense ministry" was taken. The chairman was Malenkov, and academy member A.I. Berg was named his deputy. In June 1947, the radar council became Special Committee No.3. The success of the radio-electronic industry had a decisive influence on the development of missile and space technology.

From the beginning of the Soviet Uran project, the initiative lay with the physicists. On the recommendation of academy member A.F. Joffe, in 1942, I.V. Kurchatov took over scientific leadership of the project.

Lavrenti Pavlovich Beria.

In 1943, the NKVD, the Soviet secret service, received a copy of a British special report about the possibilities of atomic weapons (the Maud Committee report). As a result, despite the lack of resources during the war, Stalin ordered a Soviet atomic weapons program started.

The Soviet atomic bomb project was initially given a relatively low priority, until information provided by spy Klaus Fuchs and the destruction of Hiroshima and Nagasaki by atomic bombs diverted Stalin's attention to the atom bomb.

At the start of organization of the atomic project, M.G. Pervukhin, the deputy of the chairman of the Council of the People's Commissariat, was placed in charge. He was simultaneously People's Commissar of the Chemical Industry. It later became apparent that the costs and scale of the work would require fresh sacrifices by the half-starved people and the nation, which had not yet recovered from the devastation caused by the war. It was also necessary, following the American example, to maintain the strictest secrecy. Only the authorities of the almighty Beria could guarantee such a level of secrecy. Committee No.1 of the State Committee for Defense was formed and Beria was named chairman of the committee.

At different times M.G. Pervukhin and B.L. Vanikov were deputies of the chairman of the atomic committee. In addition to all of his other advantages over the usual ministers, Beria had an army of unpaid workers—the prisoners of the "Gulag Archipelago"—and the many thousand-man army of the NKVD. Committee No.1 later became the 1st Main Directorate

in the Soviet Council of Ministers. Boris Lvovich Vannikov was named chairman of the 1st Main Directorate.

The realization of the atomic project required the important raw material uranium. The search for and mining of uranium by the Soviet Union was directed mainly in the Erz Mountains in the Soviet occupation zone in Germany and in Czechoslovakia. Until 1945, it was assumed that only modest quantities of uranium were present there. In June–July 1945, Soviet geologists began looking for uranium in Oberschlema, and in August the search was expanded to Schneeberg.

By order of the Soviet military commander of Schneeberg, in September 1945, the mining of bismuth, cobalt, and nickel was resumed in the Schneeberger area. Work began with a crew of ninety-three men in the area of the Weißer Hirsch mine and at the Beutschacht.

On September 14, 1945, the 9th Administration of the Ministry of the Interior formed a geological group. For a period of two months it was to search for uranium deposits in the Erz Mountains in Saxony. The work was subsequently continued by the Saxon Ore Exploration Group. On April 4, 1946, the Saxon Ore Exploration Group became the Saxon Extraction and Exploration Group.

In spring 1946, exploration for uranium ore began at the Marx-Semler mine in Oberschlema. The spa facility in Schlema was closed to civilians by order of the Soviets on November 15, 1946, because the area of the radium bath was now in the sealed-off area of the mine.

A decision by the Council of Ministers of the USSR on July 29, 1946, resulted in the Saxon Extraction and Exploration Group becoming the Saxon Mining Administration, predecessor of the later general directorate of the Wismut AG, whose first director general was Soviet Maj. Gen. Mikhail Mitrofanovich Meltsev. The administration had the Soviet Army postal number 27304.

The mining of uranium ore that now began surpassed anything known before in its scale and effects. The exploration program by the SAG/SDAG Wismut was later expanded to include all of East Germany. This resulted in the discovery of black shale deposits near Ronneburg in East Thuringia, the uranium mineralization of the coal of the

Döhlener Basin, and the Königstein sandstone beds in Saxon, Switzerland.

Production reached its peak in the 1960s, at about 7,716 tons per year. Up to and including 1990, the Wismut AG/SDAG produced 268,300 tons of uranium. The German Democratic Republic thus delivered about one third of all the uranium produced in the Soviet sphere of influence until 1990. The Soviet Union produced about 439,800 tons of uranium by 1990.

Stalin ordered Kurchatov to produce a bomb by 1948. The entire project was then moved to the city of Sarov in the Gorki Oblast (now the Oblast of Nizhni Novgorod) and renamed Arsamas-16.

Kurchatov set up the first experimental reactor on Moscow in November 1946, and it became critical just one month later. As we now know, the Soviet physicists largely designed their first atomic

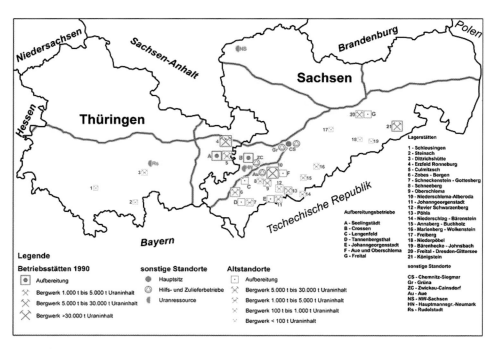

Wismut locations.

Igor Vasilyevich Kurchatov.

bomb in keeping with the information provided them by Fuchs.

By 1947, Kurchatov was already a powerful scientific leader of the project. He described himself as head of the laboratory for measurement technology at the Academy of Science (LIPAN).

In August 1947, by order of the Council of Ministers of the USSR and central committee of the Communist Party, Object 905 for the testing of atomic devices. Over the years the test site's name changed several times. Lieutenant General of Artillery Peter Mikhailovich Roshanovich (February–September 1948) was named its first commander. His successor in that position was Major General of Artillery Sergey Georgievich Kolesnikov.

The atomic test range at Semipalatinsk is about 250 miles east of the capital city of Astana, west-southwest of the city of Semei (Russian Semipalatinsk) in the Kazakh steppe at an elevation of 328 to 984 feet with mountains up to 3,900 feet high. Its position on the continent means that there are great differences between summer (up to 113° F) and winter (as low as -58° F) temperatures, with a low annual precipitation of eight to twelve inches.

The construction of personnel quarters was carried out by military units. The first work began in 1947, on the bank of the Irtysh. The Soviet Army's commander of engineering troops, Marshall Mikhail Petrovich Vorobyov, proved a great help in building the testing site.

Construction of the testing area was virtually complete by the beginning of summer 1949. The construction of a military town on the shore of the Irtysh approximately seventy-two miles downriver from Semipalatinsk began: the buildings accommodating the headquarters staff of Unit 52605, the officers' residences, a hotel for arriving participants in the tests, a two-story house for the commander of the test site. The latter, however, was used to accommodate L.P. Beria and his escort.

The experimental scientific part of the testing site was built about a mile from the riverbank and surrounded by a fence. Several office buildings were built in this area and they accommodated the research laboratories (biological, radiochemical, physical measurements) and the sector for testing of military equipment.

The center of the experimental field was located approximately thirty-five miles from occupied places. The experimental field, in whose geometric center the test tower was erected, was surrounded by several rows of barbed wire and guarded by the military. The diameter of the field was approximately twelve miles.

The command post—a semi-underground room made of steel-reinforced concrete with metal vaults doors, surrounded on the outside by heaped-up earth, was set up at a distance of six miles from

General layout of the atomic weapons test site at Semipalatinsk.

the center of the experimental field. There were niches from which the explosion could be observed.

The experimental field was a circle divided into sectors. Testing technology and systems were set up in the sectors. Each sector had a purpose. For example, there were residential houses in the civilian sector. There were also sectors for testing military technology: an artillery sector, an aviation technology sector, a sector for protective structures (trenches, dugouts) and agents, rear echelon services sector where food, clothing, etc. were tested.

To determine the effects of an atomic explosion on living organisms a sector was established where animals were housed: dogs, sheep, rabbits, pigs.

In the instrument systems that were set in the field at various distances from the hypocenter were still and movie cameras which were supposed to quantitatively determining the various parameters of the explosion.

The search for a location for the super-secret design bureau began in late 1945. At the end of April 1946, J. Shariton and P. Sernov found Sarov, where there had once been a monastery and now was a munitions factory. The choice of this location was confirmed as it was far away from large cities and also had an initial production infrastructure.

Work on the Soviet atomic bomb began in design bureau KB-11 in 1946. The research and production activities by KB-11 were kept under the strictest secrecy. Their nature and objectives were top-level state secrets.

In 1947, KB-11's staff stood at thirty-six scientific personnel. They came from various institutes, mainly from the Academy of Science of the USSR. Two projects were being worked on. The first was to

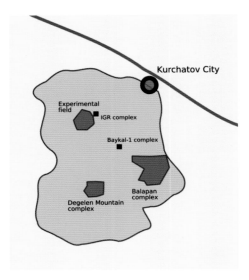

The atomic weapons test site at Semipalatinsk.

investigate the material plutonium, in the second Uranium 235. The abbreviation РДС (реактивный двигатель специальный) stood for "reactive engine special." The RDS-1 was tested.

At seven in the morning on August 29, 1949, a blinding flash lit the Semipalatinsk testing site. At the time of the explosion, where the central tower was, there appeared a glowing hemisphere, four to five times larger than the sun.

Research into the destructive effects of the nuclear explosion on military technology and military and civilian structures was carried out by specialists of the Soviet military.

In the early years of the missile industry, Korolev, whose accomplishments are often compared to those of Kurchatov by historical journalists, was not

Specification of the RDS-1, The First Soviet Atomic Bomb	
Product 501, Atomic Charge 1-200	
Explosive power	22 kilotons
Length	12.14 ft
Diameter	4.9 ft
Weight	10,141 lbs
Fissionable material	plutonium

The RDA-1, the first Soviet atomic bomb.

on the same level with respect to power and opportunities. The same was true of the material security of his laboratories and the living conditions of the scientists and specialists. Compared to those involved in nuclear energy, the raketchikis were the "poor relations." Until the 1990s, the closed "atomic cities" offered architecture, living comforts, social, cultural, and medical facilities, and consumer goods that were far beyond anything available to the employees and workers of the missile industry.

When shared work between the atomic and missile industries began in 1952, the missile experts learned, not without envy, of the unlimited production and research capabilities and the comparatively sumptuous living quarters and material goods available to the atomic industry in that time of severe shortages.

This being left behind made Korolev sick, and he frequently complained to Ustinov that he had a false appreciation for the work of the missile experts. Now, after many years, one can understand that the matter did not depend on Ustinov. The nation was simply not in the position to create comfortable conditions for all three branches of industry working in this field—the nuclear, missile, and electronic research industries.

While a transportable Soviet atomic bomb did exist in 1952, the goal of directly threatening American territory had not yet been achieved. The new missile had to be capable of transporting the RDS-1 bomb, which was twelve feet long, 4.9 feet in diameter, and weighed 10,140 pounds.

Phases of the atomic explosion on August 29, 1949.

APPENDICES

Appendix 1
A-4 Rocket Launches from Kapustin Yar, October 26 to November 13, 1947

October 18, 1947, 0747 GMT – Launch Complex A-4, booster rocket: A-4, configuration A-4 010 T.
OKB: Korolev. Altitude: 53 miles after 33 miles.
Summary: distance flown 128 miles, destroyed during the ballistic part of flight.

October 20, 1947, 0814 GMT – Launch Complex A-4, booster rocket: A-4, configuration A-4 04 T.
OKB: Korolev. Altitude: 52.8 miles after 31.5 miles.
Summary: distance flown 144 miles, deviation 108 miles left of planned target.

October 23, 1947, 1405 GMT – Launch Complex A-4, booster rocket: A-4, configuration A-4 08 T.
Failure: Payload destroyed, resulting in disintegration of rocket.
OKB: Korolev. Altitude: 8.4 miles after 8 minutes.
Summary: distance flown 18.25 miles.

October 28, 1347, 0814 GMT – Launch Complex A-4, booster rocket: A-4, configuration A-4 03 T.
OKB: Korolev. Altitude: 54 miles after 33 miles.
Summary: distance flown 173 miles, target reached.

October 31, 1341, 0814 GMT – Launch Complex A-4, booster rocket: A-4, configuration A-4 06 T.
Failure: loss of longitudinal control.
OKB: Korolev. Altitude: 0 miles.
Summary: distance flown 2 miles.

November 2, 1947, 1514 GMT – Launch complex A-4, booster rocket A-4, configuration A-4 14 N.
OKB: Korolev. Altitude: 88 miles after 33 miles.
Summary: distance flown 161.5 miles; target reached.

November 3, 1947, 1205 GMT – Launch complex A-4, booster rocket A-4, configuration A-4 30 N.
Failure: rolled after launch, loss of stabilization.

OKB: Korolev. Altitude: 0 miles.
Summary: distance flown 1.4 miles.

November 4, 1947, 1502 GMT – Launch complex A-4, booster rocket A-4, configuration A-4 01 T.
OKB: Korolev. Altitude: 55 miles after 34 miles.
Summary: distance flown 166.5 miles; target reached.

November 10, 1947, 0939 GMT – Launch complex A-4, booster rocket A-4, configuration A-4 21 N.
Failure: guidance failed.
OKB: Korolev. Altitude: 6.6 miles after 3.5 miles.
Summary: distance flown 15 miles.

November 13, 1947, 0830 GMT – Launch complex A-4, booster rocket A-4, configuration A-4 22 N.
OKB: Korolev. Altitude: 55 miles after 34 miles.

November 13, 1947, 1400 GMT – Launch complex A-4, booster rocket A-4, configuration A-4 19 N.
OKB: Korolev. Altitude: 55 miles after 34 miles.
Summary: distance flown 170.25 miles; broke up during reentry into the atmosphere.

Appendix 2
R-1 Missile Launches from Kapustin Yar, September 17 to November 5, 1948.

September 17, 1948: launch complex A-4, R-1 booster rocket, configuration R-1-4.
OKB: Korolev. Altitude: 0.6 miles.
Summary: failure of management system, beginning with a rotation about the missile's longitudinal axis, deviation of 51 degrees from guideline, altitude 3,609 feet, missile came down 7.5 miles from point of launch.

October 10, 1948: launch complex A-4, R-1 booster rocket, configuration R-1-4.
OKB: Korolev. Altitude: 62 miles after 37 miles. Flew a distance of 180 miles in successful first flight.

October 11, 1948: R-1 booster rocket.
R-1 flight – Summary: First flight by R-1 carrying scientific experiments.

October 13, 1948: launch complex A-4, R-1 booster rocket, configuration R-1-9.
Test Mission: OKB Korolev. Altitude: 62 miles after 37 miles.

October 21, 1948: launch complex A-4, R-1 booster rocket, configuration R-1-6.
Test Mission: OKB Korolev. Altitude: 62 miles after 37 miles.

October 23, 1948: launch complex A-4, R-1 booster rocket, configuration R-1-10.
Test Mission: OKB Korolev. Altitude: 62 miles after 37 miles.

November 1, 1948: launch complex A-4, R-1 booster rocket, configuration R-1-3.
Test Mission: OKB Korolev. Altitude: 62 miles after 37 miles.

November 3, 1948: launch complex A-4, R-1 booster rocket, configuration R-1-12.
Test Mission: OKB Korolev. Altitude: 62 miles after 37 miles.

November 4, 1948: launch complex A-4, R-1 booster rocket, configuration R-1-2.
Test Mission: OKB Korolev. Altitude: 62 miles after 37 miles.

November 5, 1948: launch complex A-4, R-1 booster rocket, configuration R-1-11.
OKB: Korolev. Altitude: 62 miles after 37 miles.
Summary: ninth and launch of the first R-1 test series.

Appendix 3
R-1 Missile Launches from Kapustin Yar, September 10 to October 23, 1949.

September 10, 1949: launch complex R-1, R-1 booster rocket, configuration: R-1 II-1.
OKB: Korolev. Altitude: 62 miles (37 miles).
Summary: first launch of the second series – 10 prototypes and 11 prototype missiles available. A total of 20 were launched; six failures in 16 launches.

September 11, 1949: launch complex R-1, R-1 booster rocket, configuration: R-1 II-2.
Test Mission: OKB Korolev. Altitude: 62 miles after 37 miles.

September 13, 1949: launch complex R-1, R-1 booster rocket, configuration: R-1 II-11.
Test Mission: OKB Korolev. Altitude: 62 miles after 37 miles.

September 14, 1949: launch complex R-1, R-1 booster rocket, configuration: R-1 II-4.
Test Mission: OKB Korolev. Altitude: 62 miles after 37 miles.

September 17, 1949: launch complex R-1, R-1 booster rocket, configuration: R-1 II-8.
Test Mission: OKB Korolev. Altitude: 62 miles after 37 miles.

September 19, 1949: launch complex R-1, R-1 booster rocket, configuration: R-1 II-5.
Test Mission: OKB Korolev. Altitude: 62 miles after 37 miles.

September 20, 1949: launch complex R-1, R-1 booster rocket, configuration: R-1 II-9, failure.
Test Mission: OKB Korolev. Altitude: 62 miles after 37 miles.

September 23, 1949: launch complex R-1, R-1 booster rocket, configuration: R-1 II-15, failure.
Test Mission: OKB Korolev. Altitude: 62 miles after 37 miles.

September 28, 1949: launch complex R-1, R-1 booster rocket, configuration: R-1 II-10.
Test Mission: OKB Korolev. Altitude: 62 miles after 37 miles.

October 3, 1949: launch complex R-1, R-1 booster rocket, configuration: R-1 II-14.
Test Mission: OKB Korolev. Altitude: 62 miles after 37 miles.

October 8, 1949: launch complex R-1, R-1 booster rocket, configuration: R-1 II-16.
Test Mission: OKB Korolev. Altitude: 62 miles after 37 miles.

October 10, 1949: launch complex R-1, R-1 booster rocket, configuration: R-1 II-12.
Test Mission: OKB Korolev. Altitude: 62 miles after 37 miles.

October 12, 1949: launch complex R-1, R-1 booster rocket, configuration: R-1 II-7.
Test Mission: OKB Korolev. Altitude: 62 miles after 37 miles.

October 13, 1949: launch complex R-1, R-1 booster rocket, configuration: R-1 II-17.
Test Mission: OKB Korolev. Altitude: 62 miles after 37 miles.

October 13, 1949: launch complex R-1, R-1 booster rocket, configuration: R-1 II-13.
Test Mission: OKB Korolev. Altitude: 62 miles after 37 miles.

October 15, 1949: launch complex R-1, R-1 booster rocket, configuration: R-1 II-19.
Test Mission: OKB Korolev. Altitude: 62 miles after 37 miles.

October 18, 1949: launch complex R-1, R-1 booster rocket, configuration: R-1 II-23.
Test Mission: OKB Korolev. Altitude: 62 miles after 37 miles.

October 19, 1949: launch complex R-1, R-1 booster rocket, configuration: R-1 II-22.
Test Mission: OKB Korolev. Altitude: 62 miles after 37 miles.

October 22, 1949: launch complex R-1, R-1 booster rocket, configuration: R-1 II-20.
Test Mission: OKB Korolev. Altitude: 62 miles after 37 miles.

October 23, 1949: launch complex R-1, R-1 booster rocket, configuration: R-1 II-3.
Test Mission: OKB Korolev. Altitude: 62 miles after 37 miles. Summary: twentieth and last launch of the second R-1 test series.

Appendix 4
R-2 Missile Launches from Kapustin Yar, October to December 1950.

October 1, 1950: Launch complex: R-2, R-2 booster rocket. Configuration: R-2 2.
First launch of test series – OKB Korolev.
Altitude: 93 miles (56 mi).

October 1, 1950: Launch complex: R-2, R-2 booster rocket. Configuration: R-2 3.
Failure: failed to reach target.
First series test launch – OKB Korolev.
Altitude: 93 miles (56 mi). Summary: first full-spectrum launch.

October 21, 1950: Launch complex: R-2, R-2 booster rocket. Configuration: R-2 1.
First series test launch – OKB Korolev.
Altitude: 93 miles (56 mi).

November 1, 1950: Launch complex: R-2, R-2 booster rocket. Configuration: R-2 4.
First series test launch – OKB Korolev.
Altitude: 93 miles (56 mi).

November 1, 1950: Launch complex: R-2, R-2 booster rocket. Configuration: R-2 8.

First series test launch — OKB Korolev.
Altitude: 93 miles (56 mi).

November 1, 1950: Launch complex: R-2, R-2 booster rocket. Configuration: R-2 8.
First series test launch — OKB Korolev.
Altitude: 93 miles (56 mi).

November 1, 1950: Launch complex: R-2, R-2 booster rocket. Configuration: R-2 5.
First series test launch — OKB Korolev.
Altitude: 93 miles (56 mi).

November 1, 1950: Launch complex: R-2, R-2 booster rocket. Configuration: R-2 6.
First series test launch — OKB Korolev.
Altitude: 93 miles (56 mi).

November 1, 1950: Launch complex: R-2, R-2 booster rocket. Configuration: R-2 7.
First series test launch — OKB Korolev.
Altitude: 93 miles (56 mi).

December 1, 1950: Launch complex: R-2, R-2 booster rocket. Configuration: R-2 10.
First series test launch — OKB Korolev.
Altitude: 93 miles (56 mi).

December 1, 1950: Launch complex: R-2, R-2 booster rocket. Configuration: R-2 11.
First series test launch — OKB Korolev.
Altitude: 93 miles (56 mi).

December 1, 1950: Launch complex: R-2, R-2 booster rocket. Configuration: R-2 09.
First series test launch — OKB Korolev.
Altitude: 93 miles (56 mi).

December 20, 1950: Launch complex: R-2, R-2 booster rocket. Configuration: R-2 12.
Failure: failed to reach target.
First series test launch — OKB Korolev.
Altitude: 93 miles (56 mi). Summary: twelfth and last launch in the prototype test series.

Appendix 5
R-2 Missile Launches from Kapustin Yar in July 1951.

July 1, 1951: Launch complex: R-2, R-2 booster rocket. Configuration: R-2 II-4.
Second series test launch — OKB Korolev.
Altitude: 93 miles (56 mi).

July 7, 1951: Launch complex: R-2, R-2 booster rocket. Configuration: R-2 II-5.
Second series test launch — OKB Korolev.
Altitude: 93 miles (56 mi).

July 7, 1951: Launch complex: R-2, R-2 booster rocket. Configuration: R-2 II-6.
Second series test launch — OKB Korolev.
Altitude: 93 miles (56 mi).

July 1, 1951: Launch complex: R-2, R-2 booster rocket. Configuration: R-2 II-3.
Second series test launch — OKB Korolev.
Altitude: 93 miles (56 mi).

July 1, 1951: Launch complex: R-2, R-2 booster rocket. Configuration: R-2 II-9.
Second series test launch — OKB Korolev.
Altitude: 93 miles (56 mi).

July 1, 1951: Launch complex: R-2, R-2 booster rocket. Configuration: R-2 II-2.
Second series test launch — OKB Korolev.
Altitude: 93 miles (56 mi).

July 1, 1951: Launch complex: R-2, R-2 booster rocket. Configuration: R-2 II-8.
Second series test launch — OKB Korolev.
Altitude: 93 miles (56 mi).

July 1, 1951: Launch complex: R-2, R-2 booster rocket. Configuration: R-2 II-12.
Second series test launch — OKB Korolev.
Altitude: 93 miles (56 mi).

July 1, 1951: Launch complex: R-2, R-2 booster rocket. Configuration: R-2 II-11.
Second series test launch – OKB Korolev.
Altitude: 93 miles (56 mi).

July 1, 1951: Launch complex: R-2, R-2 booster rocket. Configuration: R-2 II-10.
Second series test launch – OKB Korolev.
Altitude: 93 miles (56 mi).

July 1, 1951: Launch complex: R-2, R-2 booster rocket. Configuration: R-2 II-7.
Second series test launch – OKB Korolev.
Altitude: 93 miles (56 mi).

July 2, 1951: Launch complex: R-2, R-2 booster rocket. Configuration: R-2 II-1.
Second series test launch – OKB Korolev.
Altitude: 93 miles (56 mi).

July 27, 1951: Launch complex: R-2, R-2 booster rocket. Configuration: R-2 II-13.
Second series test launch – OKB Korolev.
Altitude: 93 miles (56 mi).
Summary: thirteenth and last launch of the second series. 12 of 13 reached their targets.

Appendix 6
R-2 Missile Launches from Kapustin Yar in August–September 1952.

August 1, 1952: Launch complex: R-2, R-2 booster rocket. Configuration: R-2-KT-6.
Third series controlled test launch – OKB Korolev.
Altitude: 93 miles (56 mi).

August 1, 1952: Launch complex: R-2, R-2 booster rocket. Configuration: R-2-KT-7.
Third series controlled test launch – OKB Korolev.
Altitude: 93 miles (56 mi).

August 1, 1952: Launch complex: R-2, R-2 booster rocket. Configuration: R-2-KT-4.
Third series controlled test launch – OKB Korolev.
Altitude: 93 miles (56 mi).

August 1, 1952: Launch complex: R-2, R-2 booster rocket. Configuration: R-2-KT-3.
Third series controlled test launch – OKB Korolev.
Altitude: 93 miles (56 mi).

August 1, 1952: Launch complex: R-2, R-2 booster rocket. Configuration: R-2-KT-2.
Third series controlled test launch – OKB Korolev.
Altitude: 93 miles (56 mi).

August 1, 1952: Launch complex: R-2, R-2 booster rocket. Configuration: R-2-KT-5.
Third series controlled test launch – OKB Korolev.
Altitude: 93 miles (56 mi).

August 1, 1952: Launch complex: R-2, R-2 booster rocket. Configuration: R-2-KT-8.
Third series controlled test launch – OKB Korolev.
Altitude: 93 miles (56 mi).

August 8, 1952: Launch complex: R-2, R-2 booster rocket. Configuration: R-2-KT-1.
Third series controlled test launch – OKB Korolev.
Altitude: 93 miles (56 mi).
Summary: first launch of the production quality-assurance series.

August 9, 1952: Launch complex: R-2, R-2 booster rocket. Configuration: R-2-KT-12.
Third series controlled test launch – OKB Korolev.
Altitude: 93 miles (56 mi).

August 9, 1952: Launch complex: R-2, R-2 booster rocket. Configuration: R-2-KT-13.
Third series controlled test launch – OKB Korolev.
Altitude: 93 miles (56 mi).

August 9, 1952: Launch complex: R-2, R-2 booster rocket. Configuration: R-2-KT-10.
Third series controlled test launch – OKB Korolev.
Altitude: 93 miles (56 mi).

August 9, 1952: Launch complex: R-2, R-2 booster rocket. Configuration: R-2-KT-9.

Third series controlled test launch – OKB Korolev.
Altitude: 93 miles (56 mi).

August 9, 1952: Launch complex: R-2, R-2 booster
rocket. Configuration: R-2-KT-11.
Third series controlled test launch – OKB Korolev.
Altitude: 93 miles (56 mi).

September 18, 1952: Launch complex: R-2, R-2
booster rocket. Configuration: R-2-KT-14.
Third series controlled test launch – OKB Korolev.
Altitude: 93 miles (56 mi).
Summary: fourteenth and last launch in the quality-
assurance test. Twelve of fourteen reached their
targets,

Appendix 7
R-1 Missile Launches from Kapustin Yar,
September to November 1948

These images are from the film Factory Testing of the R-1 Missile, 1948. They show launch preparations, the launch itself, and the flight to the target area and the impact crater.

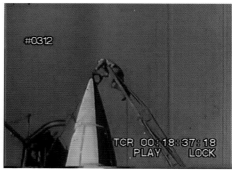

Failed launch

Successful launch on
October 10, 1948

#0312
TCR 00:30:36:06
PLAY   LOCK

#0312
TCR 00:30:38:10
PLAY   LOCK

#0312
TCR 00:30:41:11
PLAY   LOCK

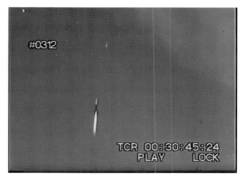

#0312
TCR 00:30:45:24
PLAY   LOCK

#0312
TCR 00:30:49:02
PLAY   LOCK

#0312
TCR 00:30:58:08
PLAY   LOCK

127

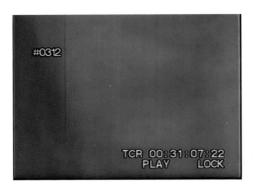

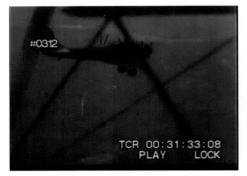

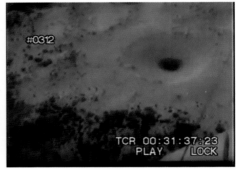

View of the impact crater.